A TEXT BOOK OF

ENGINEERING PHYSICS-II

(BASIC PHYSICS)

SEMESTER – II
FIRST YEAR DIPLOMA COURSES IN ENGINEERING AND TECHNOLOGY

AS PER SBTE'S NEW REVISED SYLLABUS
FOR JHARKHAND UNIVERSITY, 2018

Dr. M. S. PAWAR
M. Sc. and D.C.S., MBA
Department of Physics,
Pimpri-Chinchwad Educational Trust's
Pimpri-Chinchwad Polytechnic,
Pradhikaran, Nigadi, PUNE – 411 044.

•

Dr. M. A. SUTAR
M. Sc., MCM, P.G.D.O.M.
Department of Physics,
Dr. D. Y. Patil Pratishthan's
Y. B. Patil Polytechnic,
Akurdi, PUNE 411 044.

N4411

F.Y. DIPLOMA (SEMESTER – II) ENGINEERING PHYSICS - II ISBN 978-93-87686-16-8

Second Edition : **January 2019**

© : **Authors**

Published By :

NIRALI PRAKASHAN

Abhyudaya Pragati, 1312, Shivaji Nagar

Off J.M. Road, Pune – 411005

Tel - (020) 25512336/37/39, Fax - (020) 25511379

Email : niralipune@pragationline.com

➢ DISTRIBUTION CENTRES

PUNE

Nirali Prakashan : 119, Budhwar Peth, Jogeshwari Mandir Lane, Pune 411002, Maharashtra
(For orders within Pune) Tel : (020) 2445 2044, 66022708, Fax : (020) 2445 1538; Mobile : 9657703145
Email : niralilocal@pragationline.com

Nirali Prakashan : S. No. 28/27, Dhayari, Near Asian College Pune 411041
(For orders outside Pune) Tel : (020) 24690204 Fax : (020) 24690316; Mobile : 9657703143
Email : bookorder@pragationline.com

MUMBAI

Nirali Prakashan : 385, S.V.P. Road, Rasdhara Co-op. Hsg. Society Ltd.,
Girgaum, Mumbai 400004, Maharashtra; Mobile : 9320129587
Tel : (022) 2385 6339 / 2386 9976, Fax : (022) 2386 9976
Email : niralimumbai@pragationline.com

➢ DISTRIBUTION BRANCHES

JALGAON

Nirali Prakashan : 34, V. V. Golani Market, Navi Peth, Jalgaon 425001, Maharashtra,
Tel : (0257) 222 0395, Mob : 94234 91860; Email : niralijalgaon@pragationline.com

KOLHAPUR

Nirali Prakashan : New Mahadvar Road, Kedar Plaza, 1st Floor Opp. IDBI Bank, Kolhapur 416 012
Maharashtra. Mob : 9850046155; Email : niralikolhapur@pragationline.com

NAGPUR

Nirali Prakashan : Above Maratha Mandir, Shop No. 3, First Floor,
Rani Jhanshi Square, Sitabuldi, Nagpur 440012, Maharashtra
Tel : (0712) 254 7129; Email : niralinagpur@pragationline.com

DELHI

Nirali Prakashan : 4593/15, Basement, Agarwal Lane, Ansari Road, Daryaganj
Near Times of India Building, New Delhi 110002 Mob : 08505972553
Email : niralidelhi@pragationline.com

BENGALURU

Nirali Prakashan : Maitri Ground Floor, Jaya Apartments, No. 99, 6th Cross, 6th Main,
Malleswaram, Bengaluru 560003, Karnataka; Mob : 9449043034
Email: niralibangalore@pragationline.com

Other Branches : Hyderabad, Chennai

niralipune@pragationline.com | www.pragationline.com

Also find us on ⨍ www.facebook.com/niralibooks

Preface

It gives us a sense of satisfaction to bring out this New revised edition of **Engineering Physics - II**.

This book aims at providing a complete coverage of the needs of First Year students as per S.B.T.E's. revised syllabus. The entire revised syllabus has been covered keeping in view the non-availability of the complete subject matter through a single source.

The difficult articles have been explained in a simple language providing, wherever necessary, neat and well explained diagrams so that even an average student may be able to follow it independently. A sufficient number of solved examples and problems with answers are given at the end of each topic. Formulae specifying symbol meaning are enlisted before solving the examples. Synopsis is given at the starting of each topic which will give a fair idea about the task to be studied in the chapter. Solving problem is the big problem for most of the students. In this book, attempt has been made to simplify this difficulty by solving examples in simple style and in sufficient details to follow the developments.

We are thankful to Prof. Vidya S. Byakod, Principal, Pimpri-Chinchwad Educational Trust's, Pimpri-Chinchwad Polytechnic, for giving valuable suggestions while writing this book. We are also thankful to the management of Pimpri-Chinchwad Educational Trust, especially to Late Shri. S. B. Patil (President), Shri. Dnyaneshwarji Landge (Vice President) and Shri. V. S. Kalbhor (Secretary), Shri. P. D. Patil (Director Dr. D. Y. Patil Pratishthan), Principal Hiremath Sir (Y. B. Patil Polytechnic) and our colleagues from Y. B. Patil Polytechnic and Dr. D. Y. Patil Institute of Technology, who encouraged us for the successful completion of this venture.

We are grateful to the publisher Shri. Dineshbhai Furia, Shri. Jignesh Furia and staff of Nirali Prakashan, Pune for providing all facilities required and bringing out this book in the short span of time at their disposal.

During the progress of writing this book, Prof. S. S. Deo, Prof. Handore, Prof. Shirsetti, Mrs. Rohini M. Pawar, Shri. S. D. Jagtap from P. C. Polytechnic, Nigadi. Prof. B. S. Salunkhe, from N. M. V. Polytechnic, Talegaon, Prof. More, Prof. Bhosale, Prof. Saindane, from J. N. I. O. T. Katraj extended their valuable help, co-operation and encouragement. We are really thankful for valuable corrections and suggestions to Prof. Dumne Sharadchandra Ex. H.O.D. in Lal Bahadur Shastri Sr. College in District Nanded. We are really thankful to all of them.

We sincerely hope this edition will be well received by students and teachers of all polytechnics.

The authors will appreciate suggestions from teachers and students for the improvement of the book.

We take this opportunity to extend our good wishes to the students and teachers of all polytechnics.

– Authors

Syllabus

1. LIGHT   **(Hours 3, Marks 6)**

Properties of Light

Reflection, Refraction, Snell's law, Physical significance of refractive index, Definition of dispersion of light along with ray diagram.

(Numericals on refractive index)

2. ELECTRIC FIELD AND POTENTIAL

2.1 ELECTRIC FIELD   **(Hours 5, Marks 8)**

Electric charge, Coulomb's inverse square law, Definition of unit charge, Electric field, Electric lines of force and their properties, Electric field intensity, Electric flux, Electric flux density.

(Numericals on Coulomb's law, Electrical intensity)

2.2 ELECTRIC POTENTIAL   **(Hours 5, Marks 8)**

Concept of potential, Definition and unit, Potential due to point charge using integration method, Potential difference between two points, Definition of dielectric strength and breakdown potential.

(Numericals on electric potential)

2.3 CAPACITY AND CONDENSERS   **(Hours 3, Marks 6)**

Electrostatics capacity and its S.I. unit, Capacity of parallel plate condenser, Condensers in series and parallel (Formula only, no derivation), Uses of condensers.

(Simple problems)

3. CURRENT ELECTRICITY   **(Hours 3, Marks 8)**

Ohm's law, Resistance and its unit, Specific resistance, Factors affecting resistance, Kirchhoff's law and its application to Wheatstone's bridge circuit.

4. FIBER OPTICS   **(Hours 5, Marks 8)**

Introduction, Total internal reflection, Critical angle, Acceptance angle. Structure of optical fiber, Numerical aperture, Fiber optic materials, Types of optical fibers, Applications in communication systems.

(Numericals on critical angle, numerical aperture)

5. BAND THEORY OF SOLIDS   **(Hours 5, Marks 8)**

Energy levels in solids, Valence and conduction bands, Forbidden gap, Conductors, Semiconductors and Insulators, Intrinsic and extrinsic semiconductors, P-type and N-type semiconductors, P-N junction diode - forward and reverse biased characteristics.

(No Numericals)

6. MODERN PHYSICS

6.1 PHOTOELECTRICITY   **(Hours 3, Marks 6)**

Concept of photon, Plank's hypothesis, Properties of photon, Photo electric effect, Laws of photoelectric effect, Work function, Einstein's photoelectric equation (no derivation), Basic concept of solar energy.

(Numericals on Energy of photon, work function, photoelectric equation)

6.2 LASER   **(Hours 1, Marks 4)**

Properties of laser, Characteristics and applications of laser

6.3 X-RAYS   **(Hours 2, Marks 6)**

Introduction to X-rays, Production of X-rays using Coolidge tube, minimum wavelength of X-rays, Properties and applications of X-rays.

(Numericals on minimum wavelength of X-rays)

7. INTRODUCTION TO NANOTECHNOLOGY   **(Hours 3, Marks 6)**

Definition of nanoscale, nanometer and nanoparticle, Applications of nanotechnology - electronics, automobiles, medical, textile, cosmetics, environmental, space and defence.

8. NON-CONVENTIONAL SOURCES OF ENERGY   **(Hours 4, Marks 6)**

Introduction – Non-renewable and renewable (alternate) energy sources, Examples – Solar energy, Wind energy, Tidal energy, Geo-thermal energy and Bio-mass. Advantages and disadvantages of renewable energy.

●●●

Contents

1. LIGHT   **1.1 – 1.6**

   1.1 Reflection   1.1

   1.2 Refraction   1.2

   1.3 Law of Refraction (Snell's Law) and Physical Significance of Refractive Index   1.3

   1.4 Dispersion   1.3

      Summary   1.4

      Important Formulae   1.4

      Solved Examples   1.4

      Exercise   1.6

      Problems for Practice   1.6

2. ELECTRIC FIELD   **2.1 – 2.12**

   2.1 Introduction   2.1

   2.2 Coulomb's Law (Inverse Square Law)   2.2

   2.3 Electric Field   2.3

   2.4 Intensity of Electric Field (E) Or Electric Intensity at a Point   2.3

   2.5 Electric Lines of Force   2.4

   2.6 Electric Flux (Ψ) (psi)   2.5

   2.7 Electric Flux Density (D) at a point and Relation Between E and D   2.5

   2.8 Electric field intensity due to a charged sphere   2.6

      Summary   2.6

      Important Information and Conversions   2.7

      Important Formulae   2.7

      Solved Examples   2.7

      Exercise   2.12

      Problems for Practice   2.12

3. ELECTRICAL POTENTIAL   **3.1 – 3.10**

   3.1 Introduction   3.1

   3.2 Electric Potential (Concept of Potential)   3.1

   3.3 Potential Due to a Point Charge   3.3

   3.4 P.D. between Two Points   3.4

   3.5 Potential of a Sphere or Potential Due to a Charged Sphere   3.5

   3.6 Dielectric Strength   3.6

      Summary   3.7

      Conversions and Formulae   3.7

      Solved Examples   3.7

      Exercise   3.9

      Problems for Practice   3.10

4. CAPACITY AND CONDENSERS   **4.1 – 4.16**

   4.1 Introduction   4.1

   4.2 Capacitance, Unit and Definition of 1 Farad   4.1

   4.3 Principle of a Condenser (or a Capacitor)   4.2

   4.4 Capacity of a Parallel-plate Condenser (Derivation)   4.3

4.5	Combination of Capacitances	4.4
	4.5.1 Series Combination and Expression for Effective Capacitance (Condensers in series)	4.4
	4.5.2 Parallel Combination and Expression for Effective Capacitance (Condensers in Parallel)	4.5
	Summary	4.6
	Conversions	4.6
	Important Formulae	4.6
	Solved Examples	4.7
	Exercise	4.14
	Problems for Practice	4.15

5. CURRENT ELECTRICITY   **5.1 – 5.18**

5.1	Introduction	5.1
5.2	Electric Current (I)	5.2
5.3	Resistance (R)	5.3
5.4	Resistivity or Specific Resistance (ρ)	5.3
5.5	Conductance (G)	5.4
5.6	Conductivity or Specific Conductance (σ)	5.4
5.7	Ohm's Law	5.4
5.8	Combination of Resistances	5.5
	5.8.1 Resistances in series	5.5
	5.8.2 Resistances in parallel	5.6
5.9	Kirchhoff's Laws	5.7
	5.9.1 Kirchhoff's First Law (Current Law)	5.7
	5.9.2 Kirchhoff's Second Law (Voltage Law)	5.7
5.10	Application of Kirchhoff's Laws to Wheatstone's Bridge Circuits	5.8
	Summary	5.9
	Important Formulae	5.10
	Solved Examples	5.10
	Exercise	5.17
	Problems for Practice	5.17

6. FIBER OPTICS   **6.1 – 6.14**

6.1	Optical Fiber	6.1
6.2	Principle of Optical Fiber or Total Internal Reflection (T.I.R.)	6.2
6.3	Propagation in Optical Fiber	6.3
6.4	Structure of Optical Fiber (Fiber Optic Construction)	6.4
6.5	Fiber Optic Materials	6.4
6.6	Critical angle (θ_c)	6.5
6.7	Acceptance angle (θ_a)	6.5
6.8	Numerical Aperture (N_A)	6.6
6.9	Types of Optical Fiber	6.6
	6.9.1 Single Mode Step Index Optical Fiber	6.6
	6.9.2 Multimode Step Index Optical Fiber	6.7
	6.9.3 Multimode Graded Index Optical Fiber	6.8
6.10	Transmission Characteristics of Optical Fibers	6.8
6.11	Advantages of Optical Fibers in Communication over Ordinary Cable Communication	6.9

6.12	Applications of Optical Fibers	6.9
	Summary	6.9
	Important Formulae	6.10
	Solved Examples	6.10
	Exercise	6.13
	Problems for Practice	6.14

7. BAND THEORY OF SOLIDS   7.1 – 7.12

7.1	Introduction	7.1
	7.1.1 Formation of Energy Band in Solids	7.2
7.2	Energy Band Diagrams for Conductors, Semiconductors and Insulators	7.3
	7.2.1 Conductor	7.3
	7.2.2 Semiconductors	7.4
	7.2.3 Insulator	7.4
7.3	Intrinsic and Extrinsic Semiconductors	7.5
	7.3.1 Intrinsic Semiconductor	7.5
	7.3.2 Extrinsic Semiconductor	7.6
7.4	PN Junction Diode - Forward and Reverse Biased Characteristics	7.8
	Summary	7.11
	Exercise	7.12

8. PHOTOELECTRICITY   8.1 – 8.14

8.1	Photoelectric Effect	8.1
8.2	Planck's Hypothesis (Planck's Quantum Theory)	8.2
8.3	Concept of Photon	8.2
8.4	Properties of Photons (Characteristics)	8.2
8.5	Characteristics of Photoelectric Effect	8.3
8.6	Definitions	8.3
8.7	Einstein's Photoelectric Equation	8.4
8.8	Photoelectric Cell	8.5
8.9	Basic Concept of Solar Energy	8.6
	8.9.1 Solar Energy	8.6
	8.9.2 Importance of Solar Energy	8.6
	8.9.3 Development of Solar Energy in the World	8.6
	8.9.4 Development of Solar Energy in India	8.7
	8.9.5 Uses of Solar Energy	8.7
	8.9.6 Advantages of Solar Energy	8.7
	8.9.7 Disadvantages of Solar Energy	8.8
	8.9.8 Impacts of solar energy on Environment	8.8
	Summary	8.8
	Important Formulae	8.9
	Solved Examples	8.10
	Exercise	8.13
	Problems for Practice	8.14

9.	**LASER**	**9.1 – 9.8**
9.1	Light Amplification by Stimulated Emission of Radiation (LASER)	9.1
9.2	Working Principle of Laser	9.2
9.3	Absorption or Stimulated Absorption	9.3
9.4	Spontaneous Emission	9.3
9.5	Stimulated Emission	9.3
9.6	Population Inversion	9.4
9.7	Pumping Methods	9.4
9.8	Optical Pumping (Three Energy Level Laser System)	9.5
9.9	Properties or Characteristics of Laser	9.6
9.10	Helium-Neon Laser (He-Ne Gas Laser)	9.6
9.11	Applications of Laser	9.7
	Summary	9.8
	Exercise	9.8
10.	**X-RAYS**	**10.1 – 10.8**
10.1	Introduction	10.1
10.2	Production of X-rays using Coolidge Tube	10.1
10.3	Two Types of X-rays	10.3
	10.3.1 Characteristic X-Rays (Line Spectrum)	10.3
	10.3.2 The Continuous X-Ray (Spectrum)	10.3
10.4	Properties of X-rays	10.5
10.5	Applications of X-rays	10.5
	Summary	10.6
	Important Formulae	10.6
	Solved Examples	10.6
	Exercise	10.8
	Problems for Practice	10.8
11.	**INTRODUCTION TO NANOTECHNOLOGY**	**11.1 – 11.6**
11.1	Introduction (Amazing - Interesting - Novelty (Idea About Nanotechnology)	11.1
11.2	What is Nano, Nanoscale, Nanometer, Nanoparticle, Nanostructured Material ?	11.1
11.3	Applications of Nanotechnology	11.2
	11.3.1 Applications of Nanotechnology in Electronics (Nanoelectronics)	11.3
	11.3.2 Applications in Automobiles	11.3
	11.3.3 Applications in Medical (Nanotechnology in Medicine)	11.4
	11.3.4 Nanotechnology in Textile	11.4
	11.3.5 Application in Cosmetics	11.4
	11.3.6 Environmental Applications	11.4
	11.3.7 Nanotechnology in Space and Defence	11.5
	Summary	11.5
	Exercise	11.6

12. NON-CONVENTIONAL SOURCES OF ENERGY   12.1 – 12.12

12.1	Introduction	12.1
12.2	Energy Sources	12.1
12.3	Non-Renewable Sources of Energy (Conventional Energy Sources)	12.3
12.3.1	Future of Non-Renewable Energy Sources	12.4
12.3.2	Advantages of Non-renewable Energy	12.4
12.3.3	Disadvantages of Non-renewable Energy	12.4
12.4	Renewable Sources of Energy (Non-conventional Energy Sources)	12.5
12.4.1	Necessity of Renewable Energy	12.5
12.4.2	Advantages of Renewable Energy	12.5
12.4.3	Disadvantages of Renewable Energy	12.5
12.4.4	Various Renewable Energy Sources	12.6
12.5	Solar Energy	12.6
12.5.1	Importance of Solar Energy	12.6
12.5.2	Uses of Solar Energy	12.6
12.5.3	Advantages of Solar Energy	12.7
12.5.4	Disadvantages of Solar Energy	12.7
12.5.5	Impacts of solar energy on Environment	12.7
12.6	Wind Energy	12.8
12.6.1	Advantages of Wind Energy	12.8
12.6.2	Disadvantages of Wind Energy	12.8
12.7	Tidal Energy	12.8
12.7.1	Advantages of Tidal Energy	12.9
12.7.2	Disadvantages of Tidal Energy	12.9
12.8	Geothermal Energy	12.10
12.8.1	Advantages of Geothermal Energy	12.10
12.8.2	Disadvantages of Geothermal Energy	12.10
12.9	Biomass Energy	12.11
12.9.1	Advantages of Biomass Energy	12.11
12.9.2	Disadvantages of Biomass Energy	12.12
	Summary	12.12
	Exercise	12.12

University Question Paper : Nov. 2018   **P.1 – P.2**

●●●

1

CHAPTER

LIGHT

1.1   Reflection

1.2   Refraction

1.3   Law of Refraction (Snell's Law) and Physical Significance of Refractive Index

1.4   Dispersion

Summary

Important Formulae

Solved Examples

Exercise

Problems for Practice

1.1 REFLECTION

When a ray of light is allowed to fall on a reflecting surface such as mirror, it is observed that light reflects in the same medium as shown in Fig. 1.1.

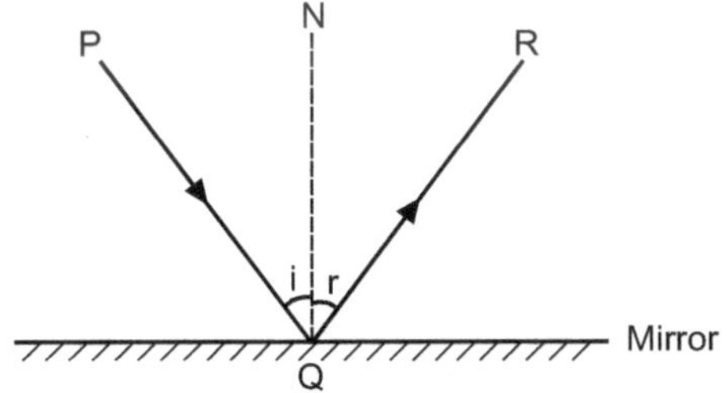

PQ = Incident ray

QR = Reflected ray

NQ = Normal to reflecting surface

Fig. 1.1

Angle made by incident ray with normal is called angle of incidence i.e. $\angle$ PQN.

Angle made by reflected ray with normal is called angle of reflection i.e. $\angle$ NQR.

Laws of reflection :

(1)   The angle of incidence is equal to the angle of reflection, i = r.

(2)   The incident ray, normal and reflected ray, lie in one plane.

If angle of incidence is zero i.e. incident light is along the normal, then reflected light returns along the same path in which light is incident.

Instead of PQ if QR becomes incident ray, then PQ becomes reflected ray, this is the principle of reversibility of light.

1.2 REFRACTION

We know that in a homogeneous medium, a ray of light travels in a straight line. But when a ray of light is incident on a plane transparent medium such as glass, it is observed that in the second medium, the ray of light deviates from its original path as shown in Fig. 1.2.

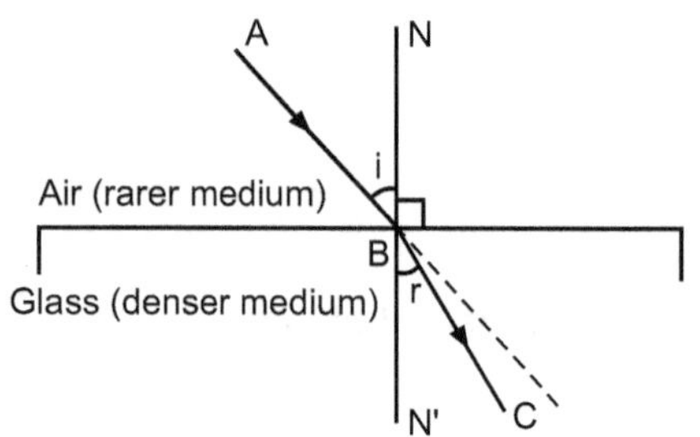

AB = Incident ray

BC = Refracted ray

NN' = Normal

$\angle$ i = Angle of incidence

$\angle$ r = Angle of refraction

Fig. 1.2

This bending of light is called as *refraction*.

The bent ray BC is called as *refracted ray*.

The refracted ray bends towards or away from the normal depending upon whether the second medium is optically denser or rarer than the first. It has been experimentally found that in the second medium, velocity of light also changes.

1.3 LAW OF REFRACTION (SNELL'S LAW)

AND PHYSICAL SIGNIFICANCE OF REFRACTIVE INDEX   (Nov. 18)

Snell's law : For any two media, the ratio of the sine of angle of incidence to the sine of angle of refraction is constant. This law is known as Snell's law.

Thus $\quad\dfrac{\sin i}{\sin r}$ = Constant

where, i – Angle of incidence

   r – Angle of refraction.

This constant is called refractive index of second medium with respect to first and is denoted by $_1\mu_2$.

$\therefore \qquad \dfrac{\sin i}{\sin r} = {_1\mu_2} = \dfrac{\mu_2}{\mu_1}$ = Constant

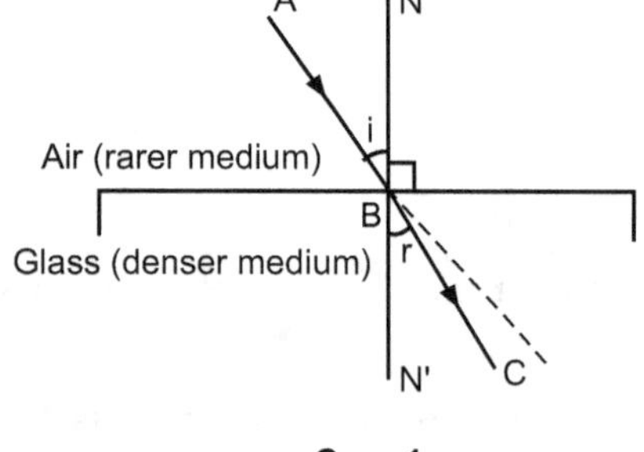

Case 1

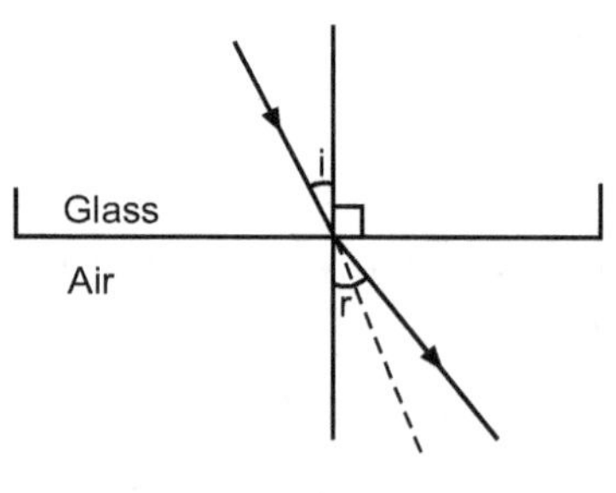

Case 2

Fig. 1.3

Case 1 : When light enters from air (rarer medium) into glass (denser medium), the ray bends towards normal. Thus, i > r.

$$\therefore \qquad \frac{\sin i}{\sin r} = \text{Constant} > 1$$

Constant is called refractive index of glass with respect to air and is denoted as $_a\mu_g$.

$$\therefore \qquad _a\mu_g = \frac{\sin i}{\sin r} > 1$$

$$\text{Also} \quad _a\mu_g = \frac{\text{Velocity of light in air}}{\text{Velocity of light in glass}}$$

Case 2 : When light enters from glass (denser medium) into air (rarer medium), the ray bends away from normal. Thus, i < r.

$$\therefore \qquad \frac{\sin i}{\sin r} = \text{Constant} < 1$$

Constant is called refractive index of air with respect to glass and is denoted as $_g\mu_a$.

$$\therefore \qquad _g\mu_a = \frac{\sin i}{\sin r} < 1$$

$$\text{Also} \quad _g\mu_a = \frac{\text{Velocity of light in glass}}{\text{Velocity of light in air}}$$

$$\text{Thus} \qquad _a\mu_g = \frac{1}{_g\mu_a}$$

1.4 DISPERSION

The light rays of each colour have certain angle of deviation of their own i.e. the angles of deviation of different rays are different.

White light is made of seven colours. When this light incident on a prism, all the seven colours have the same angle of incidence but since they have different angle of deviation they get separated when they come out of the prism.

Therefore, when white light passes through a prism, seven rays get separated. They form a spectrum of seven colours - dispersion. *The dispersion is defined as separation of constituent colours of incident light by the prism and the medium prism (glass) is called dispersive medium.*

The dispersion of white light by a prism is shown in Fig. 1.4.

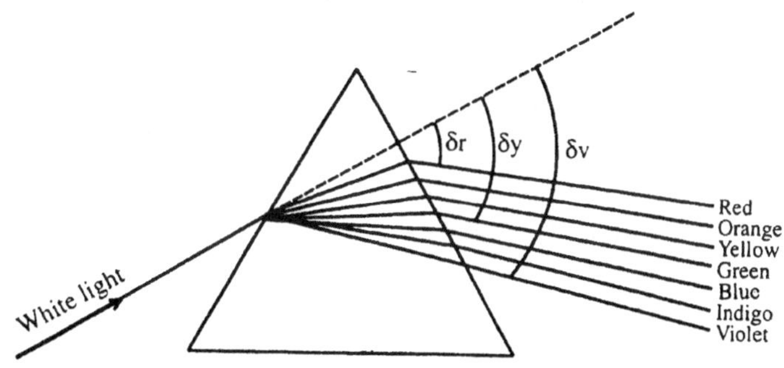

Fig. 1.4 : Dispersion of white light

SYMMARY

- When light is incident on the reflecting surface, such as mirror, it reflects back in the same medium. The angle of incidence is equal to the angle of reflection.

- When light enters from one medium into another, the ray of light bends (changes its path). This property of light is called as refraction.

- For given pair of media, ratio of sine of angle of incidence to sine of angle of refraction is always constant. This is known as Snell's law.

- When white light is incident on the prism, its constituent colours get separated. This phenomenon is called as dispersion.

IMPORTANT FORMULAE

1. $\mu = \dfrac{\sin i}{\sin r}$, where μ – Refractive index

i – Angle of incidence

r – Angle of refraction

2. $_a\mu_g = \dfrac{\text{Velocity of light in air}}{\text{Velocity of light in glass}} = \dfrac{\mu_g}{\mu_a}$

where $_a\mu_g$ - Refractive index of glass with respect to air

or $_1\mu_2 = \dfrac{\text{Velocity of light in medium 1}}{\text{Velocity of light in medium 2}} = \dfrac{\mu_2}{\mu_1}$

SOLVED EXAMPLES

Example 1.1 : *An equilateral glass prism under minimum deviation condition has an angle of incidence 40°. Find it's refractive index.*

Solution : $i = 40°$, $A = 60°$ ($\because$ prism is equilateral), $\mu = ?$

We have, $\mu = \dfrac{\sin i}{\sin r}$

$\mu = \dfrac{\sin i}{\sin (A/2)} = \dfrac{\sin (40)}{\sin (60/2)} = \dfrac{\sin (40)}{\sin (30)}$

$$\boxed{\mu = 1.285}$$

Example 1.2 : *The refractive index of glass with respect to air is 1.51 and velocity of light in air is 3×10^8 m/s. Calculate the velocity of light in glass.*

Solution : Given : $_a\mu_g = 1.51$

$v_{air} = 3 \times 10^8 \, m/s$

$v_{glass} = ?$

$$a\mu_g = \frac{\text{Velocity of light in air}}{\text{Velocity of light in glass}}$$

$$\text{Velocity of light in glass} = \frac{\text{Velocity of light in air}}{a\mu_g} = \frac{3 \times 10^8}{1.51}$$

$$\boxed{\text{Velocity of light in glass} = 1.986 \times 10^8 \ \text{m/s}}$$

Example 1.3 : *Refractive index of glass with respect to air is 1.5. Refractive index of water with respect to air is 1.2. Calculate refractive index of water with respect to glass.*

Solution : Given : Method 1 :

$$a\mu_g = 1.5$$

$$a\mu_w = 1.2$$

$$g\mu_w = \ ?$$

$$\text{i.e.} \quad g\mu_w = \frac{v_g}{v_w} = \ ?$$

We have,

$$a\mu_g = \frac{v_a}{v_g}$$

$$\therefore \quad 1.5 = \frac{v_a}{v_g} \qquad \qquad \dots (1)$$

and

$$a\mu_w = \frac{v_a}{v_w}$$

$$\therefore \quad 1.2 = \frac{v_a}{v_w} \qquad \qquad \dots (2)$$

Dividing equation (2) by equation (1),

$$\therefore \quad \frac{1.2}{1.5} = \frac{v_a/v_w}{v_a/v_g}$$

$$\therefore \quad \frac{1.2}{1.5} = \frac{v_g}{v_w}$$

$$0.8 = \frac{v_g}{v_w}$$

Thus,

$$\boxed{g\mu_w = \frac{v_g}{v_w} = 0.8}$$

OR

Method 2 :

$$a\mu_g = 1.5$$

$$a\mu_w = 1.2$$

$$g\mu_w = \ ?$$

$$a\mu_g = 1.5 = \frac{\mu_g}{\mu_a} \qquad \qquad \dots (1)$$

$$a\mu_w = 1.2 = \frac{\mu_w}{\mu_a} \qquad \qquad \dots (2)$$

Dividing equation (2) by equation (1),

$$\frac{1.2}{1.5} = \frac{\mu_w/\mu_a}{\mu_g/\mu_a}$$

$$0.8 = \frac{\mu_w}{\mu_g} = \mu_w$$

Example 1.4 : *If the broadcasting frequency of wave is 20×10^6 Hz, calculate its wavelength. (Given : Velocity of light = 3×10^8 m/s)*

Solution : Given : $\qquad n = 20 \times 10^6$ Hz

$\qquad\qquad\qquad\qquad\qquad v = 3 \times 10^8$ m/s

$\qquad\qquad\qquad\qquad\qquad \lambda = ?$

We have, $\qquad\qquad\qquad v = n\lambda$

$\therefore \qquad\qquad\qquad\qquad \lambda = \dfrac{v}{n} = \dfrac{3 \times 10^8}{20 \times 10^6}$

$$\boxed{\lambda = 15 \text{ m}}$$

EXERCISE

1.   What is refraction of light ? State laws of refraction.

2.   State Snell's law of refraction.

3.   State the principle of superposition of waves.

4.   What is interference of light ?

5.   What is constructive and destructive interference ?

6.   State conditions for constructive and destructive interference.

7.   Define diffraction.

8.   Define refraction, dispersion, diffraction.

PROBLEMS FOR PRACTICE

1.   Calculate the velocity of light in glass of refractive index 1.5, if velocity of light in air is 3×10^8 m/s.

   Ans. $v_g = 2 \times 10^8$ m/s.

2.   Refractive index of glass with respect to air is 1.5. Refractive index of water with respect to air is 1.2. Calculate refractive index of glass with respect to water.

   Ans. $_w\mu_g = 1.25$

❑❑❑

2

CHAPTER

ELECTRIC FIELD

2.1   Introduction

2.2   Coulomb's Law (Inverse Square Law)

2.3   Electric Field

2.4   Intensity of Electric Field or Electric Intensity at a Point

2.5   Electric Lines of Force

2.6   Electric Flux (ψ) (psi)

2.7   Electric Flux Density (D) at a Point and Relation between E and D

2.8   Electric Field Intensity due to a Charged Sphere

       Summary

       Important Information and Conversions

       Important Formulae

       Solved Examples

       Exercise

       Problems for Practice

2.1 INTRODUCTION

- Everyone is familiar with the fact that when dry hair is combed with a comb, crackling sound is produced. This is because comb is strongly electrified due to *friction*. In general, when a body is rubbed with a different body, electric charge is produced.

- The modern concept of electrification is based on the fact that under normal condition, a matter consists of equal number of positive and negative charges. When two bodies are rubbed together, some amount of charge is transferred from one body to another and this results in disturbing the electrical neutrality of matter.

- An atom consists of protons, neutrons and electrons. Electrons in the atom can be taken away from it. In such a case, the atom has more protons than electrons. This atom becomes positively charged. The electrons thus removed may be added to some other atom. In such a atom, there are more electrons than protons. This atom becomes negatively charged.

- If electrons are removed from atoms of a body, then the body becomes positively charged, and if electrons are added to a body, then the body becomes negatively charged.

(2.1)

- If less number of electrons are removed from a body, it is less positive and if more number of electrons are removed from a body, it becomes more positive. It is a common experience that a glass rod rubbed against a silk cloth gets a positive charge, while an ebonite rod rubbed against a piece of wool gets a negative charge. This happens due to transfer of electrons.

- It is observed that there is a force of repulsion between two similar charges and there is a force of attraction between two dissimilar charges.

2.2 COULOMB'S LAW (INVERSE SQUARE LAW)

- The scientist Charles Augustin Coulomb stated, in the year 1785, a law giving the magnitude only of the force of attraction or repulsion between two charges.

- It states that *the force of attraction or repulsion between two electric charges is directly proportional to the product of the strength of the two charges and inversely proportional to the square of the distance between them.*

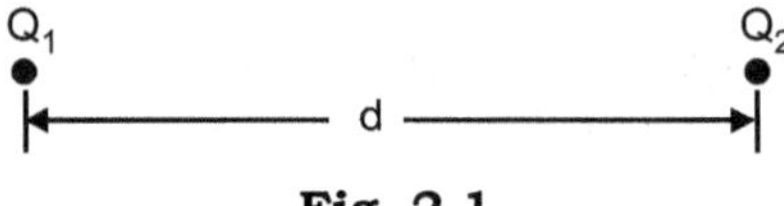

Fig. 2.1

Let Q_1, Q_2 - strengths of two charges

d - distance between them.

Then according to Coulomb's law,

$$F \propto Q_1 Q_2$$

and

$$F \propto \frac{1}{d^2}$$

Combining,

$$F \propto \frac{Q_1 Q_2}{d^2}$$

$\therefore$

$$F = \text{Constant} \times \frac{Q_1 Q_2}{d^2}$$

$$F = \frac{1}{4\pi\varepsilon} \times \frac{Q_1 Q_2}{d^2}$$

where ε (epsilon) = permittivity of the medium

but $\varepsilon = \varepsilon_o k$

$\therefore$

$$\boxed{F = \frac{1}{4\pi\varepsilon_o k} \cdot \frac{Q_1 Q_2}{d^2}}$$

where ε_o = permittivity of free space

$$= 8.85 \times 10^{-12} \, C^2/N\text{-}m^2$$

k = dielectric constant of the medium or relative permittivity of the medium.

If we substitute value of ε_o, then value of $\dfrac{1}{4\pi\varepsilon_o} \approx 9 \times 10^9$

$\therefore$

$$\boxed{F = 9 \times 10^9 \, \frac{Q_1 Q_2}{kd^2}}$$

Value of k $\approx$ 1 for air medium.

$\therefore$

$$F = 9 \times 10^9 \, \frac{Q_1 Q_2}{d^2} \quad \text{in air medium.}$$

Unit charge (or Charge of 1 coulomb) :

We have,
$$F = 9 \times 10^9 \frac{Q_1 Q_2}{kd^2}$$

If $\quad Q_1 = Q_2 = Q$ say

$$k = 1 \text{ (air medium)}$$
$$d = 1 \text{ m}$$

and $\quad F = 9 \times 10^9$ N

then,
$$9 \times 10^9 = 9 \times 10^9 \frac{Q \times Q}{1 \times 1}$$

i.e. $\quad Q^2 = 1$

$\therefore \quad Q = \pm 1$ coulomb

1 coulomb : Thus, if two equal strength charges are placed in air 1 m apart from each other and if they exert a force of 9×10^9 N on each other, then each charge is said to be a unit charge or charge of 1 coulomb.

2.3 ELECTRIC FIELD

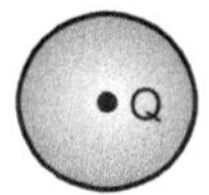 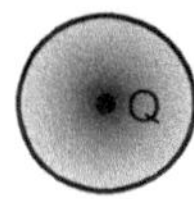

Fig. 2.2

- There is a space around an electric charge in which electric effects such as force of attraction or repulsion due to the charge can be observed. This space is called electric field of the charge. *The electric field is defined as the space around the charge in which electric effects such as attraction or repulsion due to the charge can be observed.*

- Or *electric field of a charge is the space around the charge where force of attraction or repulsion due to a charge is present.*

2.4 INTENSITY OF ELECTRIC FIELD (E) OR ELECTRIC INTENSITY AT A POINT

- Intensity of electric field at a point due to a point charge is defined as *the force acting on a unit positive charge placed at that point.*

- If strength of charge is more, then its field will be more.

Fig. 2.3

We have,
$$F = 9 \times 10^9 \times \frac{Q_1 Q_2}{kd^2}$$

Put $Q_1 = Q$ and $Q_2 = 1$ then $F = E$

$\therefore$ Intensity at 'P' is $\boxed{E = 9 \times 10^9 \dfrac{Q}{kd^2}}$

Also intensity, $\quad E = \dfrac{1}{4\pi\varepsilon_o} \dfrac{Q}{kd^2}$

M.K.S. unit of 'E' is $\dfrac{\text{Newton}}{\text{Coulomb}} = \dfrac{N}{C}$ or $\dfrac{V}{m}$

(C.G.S. unit of 'E' is dyne/stat coulomb)

2.5 ELECTRIC LINES OF FORCE   (Nov. 18)

- If a unit positive charge is placed very close to a given positive charge Q, then the force of repulsion acts on it.

- This force pushes away the unit charge. The path of this unit charge is along the line joining positions of two charges. This path is called electric lines of force of charge + Q.

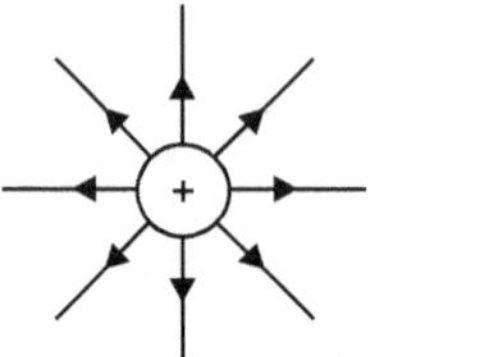

(a) Electric lines of force of a

positive charge

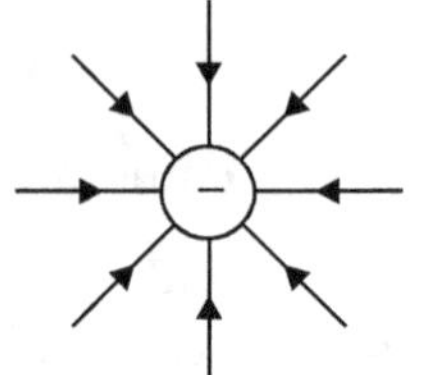

(b) Electric lines of force of a

negative charge

Fig. 2.4

- Similarly if a unit positive charge is placed on the boundary of electric field of negative electric charge –Q, then the force of attraction pulls the unit positive charge towards –Q. The path of unit charge is along the line joining positions of two charges. This path is called an electric line of force of charge –Q.

- **Thus, electric line of force of a charge is defined as the path along which a unit positive charge will move, when placed in the electric field of the charge.**

- The electric lines of force are very close to each other near the charge where the intensity of electric field is maximum. The distance between them goes on increasing as we go away from the charge i.e., as intensity of electric field goes on decreasing.

- Therefore, the number of lines of force passing through unit area is a measure of intensity of electric field.

- If the lines of force are perpendicular to the area, then maximum number of lines will pass through it and the number of lines will decrease if this area is made oblique.

Properties or Characteristics of Electric lines of force :

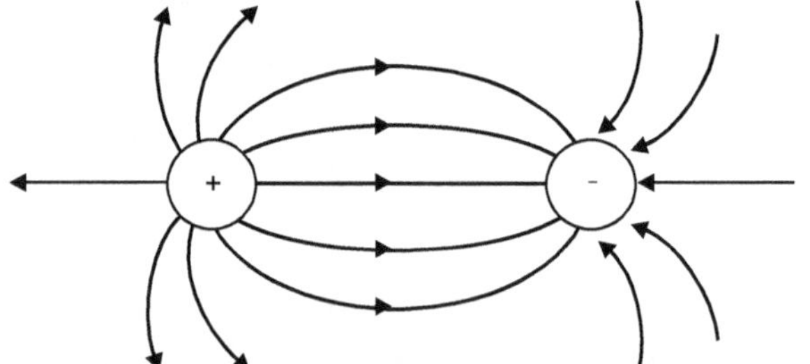

(1) Electric lines of force due to two

dissimilar charges

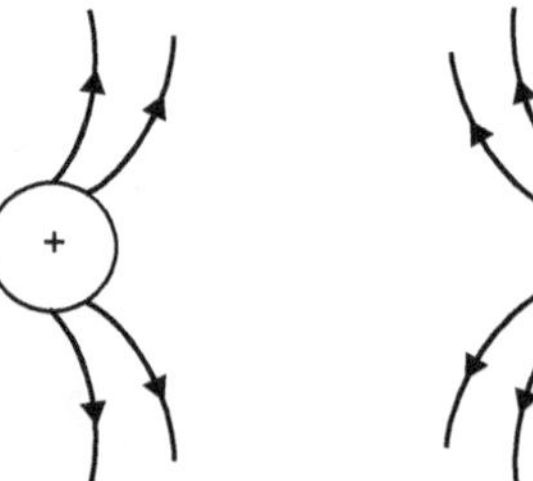

(2) Electric lines of force due to two

similar charges

Fig. 2.5

1. Electric lines of force start from positive charge and end at negative charge.

2. Electric lines of force never intersect each other. As electric intensity has only one direction at a given point, only one line passes through a given point.

3. Electric lines of force because of two charges are curved in nature.

4. The direction of electric field at a point is given by the tangent drawn to the electric line of force at that point.

5. The electric lines of force are always normal to the surface of a charged body.

6. The electric lines of force are present only on the outer surface of charged body because of force of repulsion between them. Therefore electric lines of force are not present inside the body.

7. Electric lines of force can pass through an insulator and cannot pass through conductors.

8. Electric lines of force are closer to each other in the region where intensity 'E' is large i.e. near the charge.

9. Electric lines of force are less crowded in the region where intensity 'E' is minimum.

2.6 ELECTRIC FLUX (Ψ) (PSI)

- *The total number of electric lines of force starting from a charge is called as electric flux.* It is shown by symbol ψ (psi).

- Also by definition, one coulomb of an electric charge gives rise to one electric flux.

  Hence, $\boxed{\Psi = Q}$ coulomb (C)

2.7 ELECTRIC FLUX DENSITY (D) AT A POINT AND RELATION BETWEEN E AND D

- Electric flux density is defined as *the number of electric lines of force crossing unit area held perpendicular to the electric lines of force which pass through the centre of area.*

$$D = \text{Flux density, } C/m^2$$

- Near the charge, 'D' is maximum, also 'E' is maximum, i.e. electric flux density is proportional to the electric field intensity.

- Electric field intensity E at a point is defined as the force acting on unit positive charge placed at that point.

$$D \propto E$$

and $$D = \varepsilon_o kE$$

We have, $$E = \frac{Q}{4\pi \varepsilon_o kd^2}$$

If $d = r$, then $E = \frac{1}{4\pi \varepsilon_o k} \frac{Q}{r^2}$ ... (2.1)

- Consider a point charge Q at the origin. If this is enclosed by spherical surface of radius r, then by symmetry, D due to Q have constant magnitude over the surface and is everywhere normal to the surface.

- Since total flux out of a closed surface is equal to the net charge within that surface,

$$\therefore \qquad Q = 4\pi r^2 D$$

$$\text{or} \qquad D = \frac{Q}{4\pi r^2} \qquad \qquad \dots(2.2)$$

- Comparing equations (2.1) and (2.2), we get,

$$\boxed{D = \varepsilon_o kE}$$

$$\text{or} \qquad D = \varepsilon E$$

2.8 ELECTRIC FIELD INTENSITY DUE TO A CHARGED SPHERE

- Consider a uniformly charged sphere 'A' of radius 'r'. Let Q be total amount of charge. Eventhough charges on the surface, their electric effects are as if they are all at the centre of the sphere.

- Let 'B' be the point outside the charge at a distance x. In order to find the field intensity at 'B', draw a sphere with C as a centre and x as a radius (shown dotted). The intensity E will be same at every point on the sphere B and it will be radially outward.

- The total charge enclosed by sphere 'A' and sphere 'B' is same, but the radius and hence area of sphere are different.

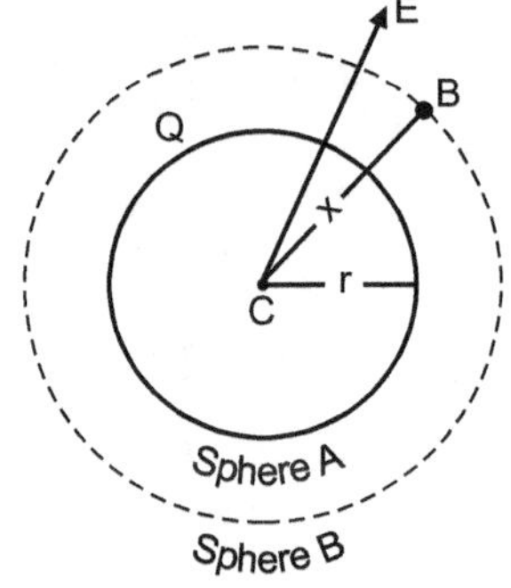

Fig. 2.6

- Now, we have

$$\text{Flux density, } D = \frac{\text{Flux}}{\text{Area}}$$

$$= \frac{\psi}{A} \qquad \qquad \text{(but } \psi = Q \text{ by Gauss's theorem)}$$

$$\therefore \qquad D = \frac{Q}{A}$$

$$D = \frac{Q}{4\pi x^2} \qquad \qquad \text{but } D = \varepsilon_0 kE$$

$$\therefore \qquad (\varepsilon_0 kE) = \frac{Q}{4\pi x^2} \qquad \therefore \qquad \boxed{E = \frac{1}{4\pi \varepsilon_0 k} \frac{Q}{x^2}}$$

SUMMARY

- Coulomb's inverse square law states that the force of attraction or repulsion between two electric charges is directly proportional to the product of their strengths and inversely proportional to the square of the distance between them.

$$F \propto \frac{Q_1 Q_2}{d^2}$$

$$F = \text{Constant} \times \frac{Q_1 Q_2}{d^2}, \text{ where Constant } = \frac{1}{4\pi \varepsilon_0 k}$$

ε_0 is called permittivity of free space and its value is 8.85×10^{-12} C^2/N–m^2

and $\quad \dfrac{1}{4\pi \varepsilon_0} \approx 9 \times 10^9$

- **Charge of 1 coulomb :** If two equal charges separated by 1 m in air, exert a force of 9×10^9 N, then each charge is said to be a charge of 1 C.

- **Electric field intensity (E) :** It is defined as the force acting on unit positive charge placed at a given point in an electric field.

- **Electric lines of force :** It is the path along which unit positive charge will move when placed in an electric field.

- **Electric flux (ψ) (psi) :** It is defined as the total number of electric lines of force originating from the given charge.

- **Electric flux density (D) :** It is defined as the number of electric lines of force passing normally through unit area.

$$D = \frac{\psi}{A}$$

IMPORTANT INFORMATION AND CONVERSIONS

1. $\varepsilon_0 = 8.85 \times 10^{-12}$ C^2 / N-m$^2 \rightarrow$ (permittivity of free space)

2. k = 1 for air medium $\rightarrow$ (dielectric constant)

3. $\dfrac{1}{4\pi\varepsilon_0} \approx 9 \times 10^9$

IMPORTANT FORMULAE

1. $F = 9 \times 10^9 \cdot \dfrac{Q_1 Q_2}{kd^2} = \dfrac{1}{4\pi\varepsilon_0} \cdot \dfrac{Q_1 Q_2}{kd^2}$

2. $E = \dfrac{\text{Force}}{\text{Charge}} = \dfrac{1}{4\pi\varepsilon_0} \dfrac{Q}{kd^2} = 9 \times 10^9 \cdot \dfrac{Q}{kd^2}$

3. $D = \varepsilon_0 \, kE$ or $D = \dfrac{\psi}{A} = \dfrac{Q}{A}$

where,

F = Force of attraction or repulsion between the charges

E = Electric intensity at a point

Q_1, Q_2 = Strength of the charges

d = Distance

ε_0 = Permittivity of free space

SOLVED EXAMPLES

Example 2.1 :

An electron is placed in an electric field of intensity 8×10^2 N/C. Calculate the force acting on the electron.

Solution :

Given : 　　　　　　　　$E = 8 \times 10^2$ N /C

Take 　　　　　　　　$e = 1.6 \times 10^{-19}$ C 　　(Charge on electron)

　　　　　　　　$F = ?$

$$\text{Electric intensity} = \frac{\text{Force}}{\text{Charge}}$$

$\therefore$　　　　　　Force　=　Electric intensity × Charge

$$= E \times e = 8 \times 10^2 \times 1.6 \times 10^{-19}$$

$$\boxed{\text{Force on electron } = 12.8 \times 10^{-17} \text{ N}}$$

Example 2.2 :

Two point charges of 16 µC and 64 µC are placed 3 m apart. Find the position of a point in between them where intensity due to two charges will be equal.

Solution :

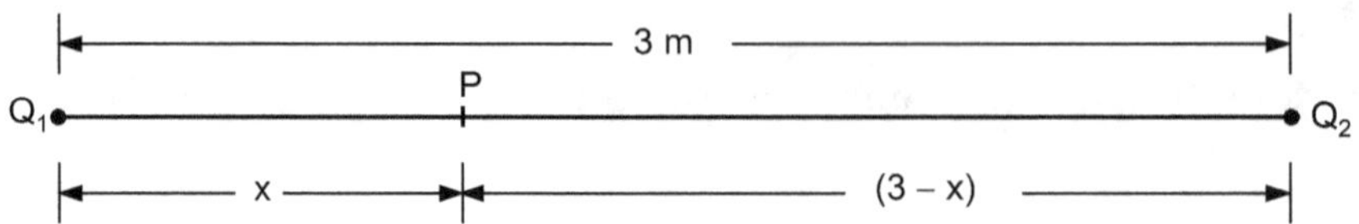

Fig. 2.7

Let 'P' be the point x metre apart from 16 µC charge where intensity due to charges is equal.

Intensity at P due to 16 µC charge is　　|　　Intensity at P due to 64 µC charge is

$$E_1 = 9 \times 10^9 \, \frac{Q_1}{kx^2} \qquad \text{... (1)} \quad \bigg| \quad E_2 = 9 \times 10^9 \, \frac{Q_2}{k\,(3-x)^2} \qquad \text{... (2)}$$

But　　　　　　　　　　$E_1 = E_2$

$\therefore$
$$9 \times 10^9 \, \frac{Q_1}{kx^2} = 9 \times 10^9 \, \frac{Q_2}{k\,(3-x)^2}$$

$\therefore$
$$\frac{Q_1}{x^2} = \frac{Q_2}{(3-x)^2}$$

$\therefore$
$$\frac{16 \times 10^{-6}}{x^2} = \frac{64 \times 10^{-6}}{(3-x)^2}$$

Taking square root on both sides, we get

$$\sqrt{\frac{16}{x^2}} = \sqrt{\frac{64}{(3-x)^2}}$$

$\therefore$
$$\frac{4}{x} = \frac{8}{(3-x)}$$

$\therefore$
$$12 - 4x = 8x$$

$\therefore$
$$\boxed{x = 1 \text{ m}}$$

Thus, the position of the point P, at which intensity is equal, is 1 m apart from 16 µC charge.

Example 2.3 :

Two charges of +16 C and –9 C respectively are placed 8 cm apart in air. Determine the position of a point at which there is no resultant electric field.

Solution :

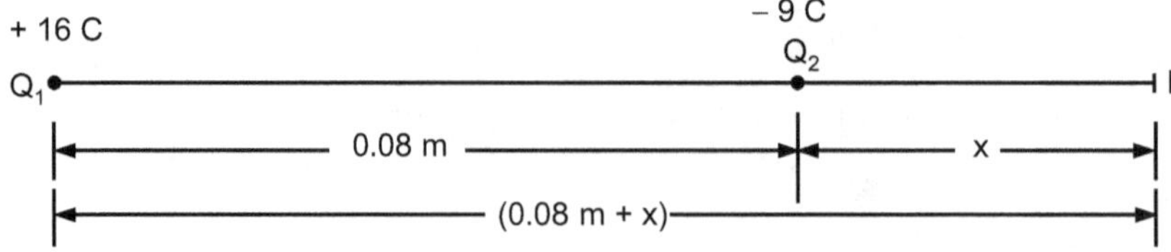

Fig. 2.8

Let 'P' be the point at which there is no resultant electric field i.e. resultant intensity is zero.

$$E_1 = 9 \times 10^9 \frac{Q_1}{k\,(0.08 + x)^2}$$

$$E_2 = 9 \times 10^9 \frac{Q_2}{kx^2}$$

Now,
$$E_1 + E_2 = 0$$

$\therefore$
$$\frac{9 \times 10^9}{k} \frac{(16)}{(0.08 + x)^2} + \frac{9 \times 10^9}{k} \frac{(-9)}{(x^2)} = 0$$

$\therefore$
$$\frac{16}{(0.08 + x)^2} - \frac{9}{x^2} = 0$$

$\therefore$
$$\frac{16}{(0.08 + x)^2} = \frac{9}{x^2}$$

Taking square root on both sides,
$$\frac{4}{(0.08 + x)} = \frac{3}{x}$$

$\therefore$
$$4x = 0.24 + 3x$$

$\therefore$
$$\boxed{x = 0.24 \text{ m}}$$

The point is 0.24 m from a charge of -9 C.

Example 2.4 :

Two like charges each of 10 μC are placed 10 cm apart in a medium of dielectric constant 2.5. Find the force between them. ($\varepsilon_0 = 8.85 \times 10^{-12}$ C^2/Nm^2)

Solution :

Given : $Q_1 = Q_2 = 10$ $\mu C = 10 \times 10^{-6}$ C, $d = 10$ cm $= 0.1$ m and $k = 2.5$

$$F = \frac{1}{4\pi\varepsilon_0 k} \frac{Q_1 Q_2}{d^2} = \frac{1}{4 \times (3.140) \times (8.85 \times 10^{-12}) \times 2.5} \times \frac{(10 \times 10^{-6}) \times (10 \times 10^{-6})}{(0.1)^2}$$

$\therefore$
$$\boxed{F = 35.96 \text{ N}}$$

Example 2.5 :

Two charges each of one coulomb are placed at a distance of 0.3 m apart in air. Calculate the force between them.

Solution :

Given : $Q_1 = Q_2 = 1$ C, $d = 0.3$ m, $k = 1$, $F = ?$

$$F = 9 \times 10^9 \frac{Q_1 Q_2}{kd^2} = 9 \times 10^9 \frac{(1) \times (1)}{(1) \times (0.3)^2}$$

$$\boxed{F = 10^{11} \text{ N}}$$

Example 2.6 :

Calculate the intensity of electric field at a point 50 cm from a charge of 4.8×10^{-6} coulomb in a medium of dielectric constant 3.6.

Solution :

Given : $d = 50$ cm $= 0.50$ m, $Q = 4.8 \times 10^{-6}$, $k = 3.6$, $E = ?$

$$E = 9 \times 10^9 \frac{Q}{kd^2} = 9 \times 10^9 \times \frac{(4.8 \times 10^{-6})}{(3.6) \times (0.5)^2}$$

$$\boxed{E = 48 \times 10^3 \text{ N/C}}$$

Example 2.7 :

Two charges of 4 μC and 16 μC are placed at 30 cm apart in air. At what point between them the electric intensity due to two charges will be equal ?

Solution :

Let 'P' be the point where electric intensity due to two charges is equal.

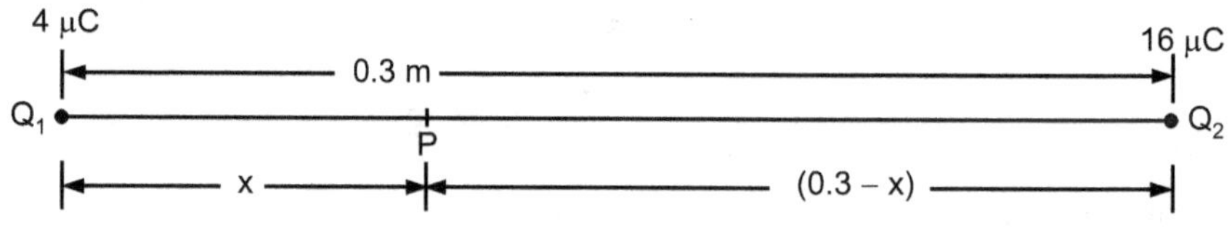

Fig. 2.9

Let intensity at 'P' due to charge 'Q_1' be 'E_1'.

Let 'P' be x metre apart from 'Q_1'.

$$E_1 = 9 \times 10^9 \frac{Q_1}{kx^2} \qquad\qquad \text{(but } k = 1 \; \because \; \text{air)}$$

$\therefore$
$$E_1 = 9 \times 10^9 \frac{Q_1}{x^2}$$

$$E_1 = 9 \times 10^9 \times \frac{(4 \times 10^{-6})}{x^2} \qquad\qquad \text{... (1)}$$

Let intensity at 'P' due to charge 'Q_2' be 'E_2'. Therefore, P is $(0.3 - x)$ metre apart from 'Q_2'.

$$E_2 = 9 \times 10^9 \frac{Q_2}{k(0.3 - x)^2} \qquad\qquad \text{(but } k = 1 \; \because \; \text{air)}$$

$$E_2 = 9 \times 10^9 \frac{Q_2}{(0.3 - x)^2}$$

$$E_2 = 9 \times 10^9 \times \frac{(16 \times 10^{-6})}{(0.3 - x)^2} \qquad\qquad \text{... (2)}$$

But at 'P', $E_1 = E_2$

Therefore, equating equation (1) and equation (2),

$$9 \times 10^9 \times \frac{(4 \times 10^{-6})}{x^2} = 9 \times 10^9 \times \frac{(16 \times 10^{-6})}{(0.3 - x)^2}$$

$$\frac{4}{x^2} = \frac{16}{(0.3 - x)^2}$$

Taking square root on both sides

$$\therefore \quad \frac{2}{x} = \frac{4}{(0.3 - x)}$$

$$\therefore \quad 2 \times (0.3 - x) = 4x$$

$$\therefore \quad \boxed{x = 0.1 \text{ m}}$$

Thus the point 'P' where resultant electric intensity due to two charges is equal is 0.1 m apart from a charge of 4 μC.

Example 2.8 :

Two electric charges of 2 μC and 16 μC are placed 30 cm apart in air. Find the position of point in between them where intensity due to two charges will be equal.

Solution :

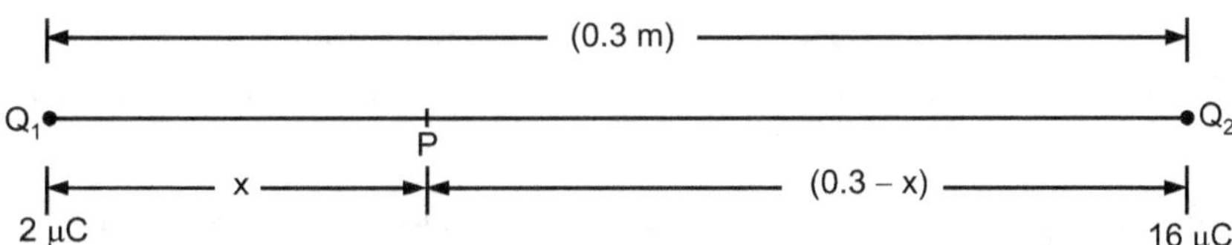

Fig. 2.10

Let 'P' be the point x metre apart from Q_1 where intensity due to two charges is equal.

Intensity at 'P' due to Q_1 is   Intensity at 'P' due to Q_2 is

$$E_1 = 9 \times 10^9 \frac{Q_1}{kx^2} \quad \ldots (1) \qquad E_2 = 9 \times 10^9 \frac{Q_2}{k(0.3 - x)^2} \quad \ldots (2)$$

But at 'P', $E_1 = E_2$.

Equating R.H.S. of equation (1) with R.H.S. of equation (2),

$$9 \times 10^9 \frac{Q_1}{kx^2} = 9 \times 10^9 \frac{Q_2}{k(0.3 - x)^2}$$

$$\frac{Q_1}{x^2} = \frac{Q_2}{(0.3 - x)^2}$$

$$\frac{2 \times 10^{-6}}{x^2} = \frac{16 \times 10^{-6}}{(0.3 - x)^2}$$

$$\frac{2}{x^2} = \frac{16}{(0.3 - x)^2}$$

Taking square root on both sides,

$$\frac{\sqrt{2}}{x} = \frac{4}{(0.3 - x)}$$

$$(1.4142) \times (0.3 - x) = 4x$$

$$(0.4243) - (1.4142 \, x) = 4x$$

$$0.4243 = 5.4142 \, x$$

$$\therefore \quad x = \frac{0.4243}{5.4142}$$

$$\therefore \quad \boxed{x = 0.0784 \text{ m}}$$

EXERCISE

1. State Coulomb's inverse square law of electrostatics and hence define one coulomb. OR Define unit charge.

2. Define the terms : Electric field, Intensity of electric field, Electric lines of force, Electric flux, Electric flux density.

3. Define electric lines of force. State properties (characteristics) of electric lines of force.

4. Obtain the relation between the flux density "D" and electric field intensity "E".

5. Define the terms : (i) Intensity of electric field, (ii) Electric flux density. State the relation between them.

PROBLEMS FOR PRACTICE

1. Two like charges when placed in a medium of dielectric constant 2.5 at a distance of 10 cm repel each other with a force of 1N. What will be the force between them when placed 5 cm apart in air ? **(Ans.** 10 N)

2. Calculate the force between two point charges of 2 coulomb each separated by a distance of 2 m in vacuum. **(Ans.** 9×10^9 N)

3. Two charges of 4 μC and 16 μC are placed 30 cm apart in air. At what point between them the electric intensity due to two charges will be equal ? **(Ans.** 10 cm from 4 μC charge.)

3

CHAPTER

ELECTRIC POTENTIAL

3.1   Introduction

3.2   Electric Potential (Concept of Potential)

3.3   Potential due to a Point Charge

3.4   P.D. between Two Points

3.5   Potential of a Sphere or Potential due to a Charged Sphere

3.6   Dielectric Strength

         Summary

         Conversions and Formulae

         Solved Examples

         Exercise

         Problems for Practice

3.1 INTRODUCTION

- The field around a charged body can be described not only by the electric field strength but also by the electric potential.

- As temperature is the deciding factor for the flow of heat from one body to another and liquid level is the deciding factor for the flow of liquid, in the similar way electric potential is the deciding factor in determining the flow of electricity.

3.2 ELECTRIC POTENTIAL (CONCEPT OF POTENTIAL)

- Consider a spring fixed at one end and let 'P' be the other end. Suppose end P is to be moved from one position to other as shown in Fig. 3.1.

- Then the external force is required to overcome the force of repulsion i.e. external work will have to be done.

- This work done in order to move point P from one position to other will be stored in the spring in the form of potential.

  i.e.   Work done = Potential of the spring

- Similarly, if a +Q charge is taken and other +1C charge is on the boundary of electric field, then there is a force of repulsion between them.

- To overcome this force and to bring +1C charge from boundary of electric field to a given charge '+Q', some work will have to be done.

  Thus,　Work done = Potential of charge

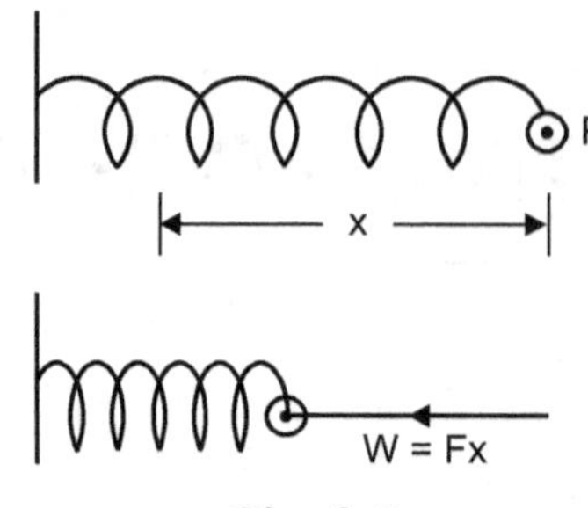

Fig. 3.1

Potential of a Charge :

- *It is the amount of work done in carrying a unit positive charge from infinity or boundary of electric field to a given charge.*

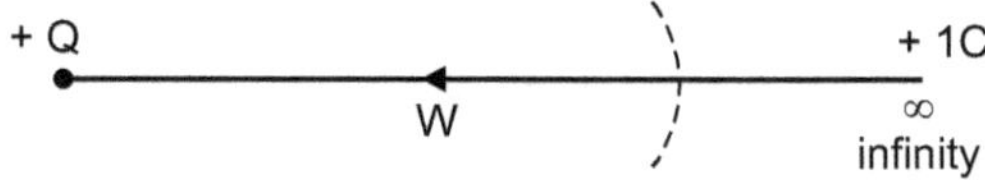

Fig. 3.2

- If the strength of the charge is more, then more work will be required to carry a charge against the field i.e. potential is more.

- In case of a negative charge, work is required in opposite direction.

- Thus, the potential due to a positive charge is taken as positive and potential due to a negative charge is taken as negative.

Potential of a Point (Absolute Potential)

- Electric potential at any point in an electric field near a charged body is defined as *the amount of work done in carrying a unit positive charge from infinity to that point against the electric field.*

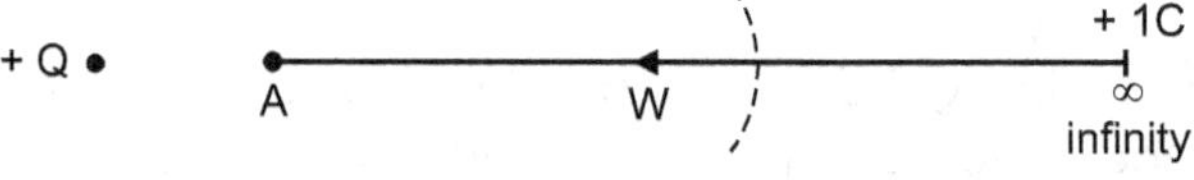

Fig. 3.3

Potential Difference Between Two Points :

- The potential difference between two points in an electric field is *the amount of work done in carrying a unit positive charge from one point to other point against the electric field.*

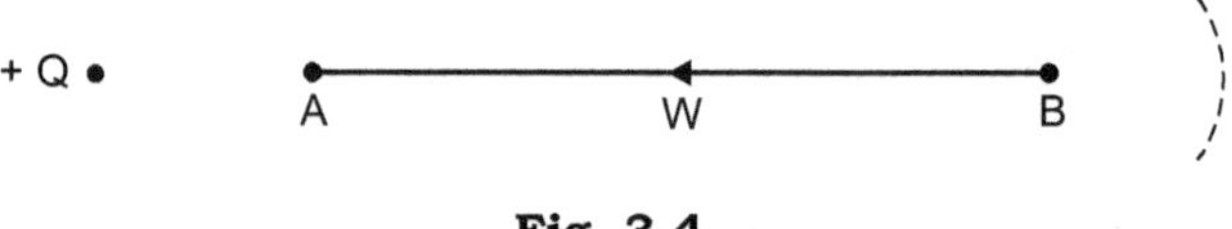

Fig. 3.4

Potential of a Sphere :

- It is the amount of work done in carrying a unit positive charge from infinity to the surface of a sphere.

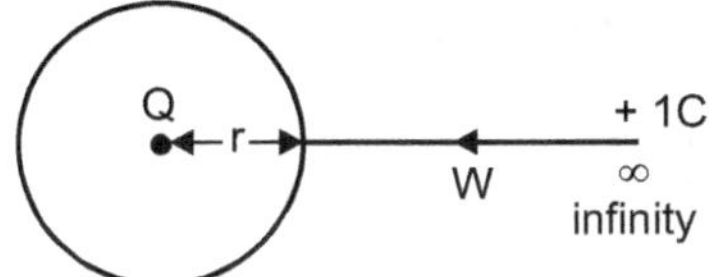

Fig. 3.5

Electric Potential (V) :

- If W is the work done in bringing a charge of small magnitude 'q' from infinity to a given point is given by,

$$\text{Electric potential, } V = \frac{\text{Work}}{\text{Charge}}$$

Potential is shown by V and its unit is volt.

$$V = \frac{W}{q} \ \dots \text{ volts}$$

$$\text{If} \quad W = 1 \text{ joule}$$
$$\text{and} \quad q = 1 \text{ coulomb}$$

$$\text{then,} \quad 1 \text{ volt} = \frac{1 \text{ joule}}{1 \text{ coulomb}}$$

- Thus, *the electric potential at a point is said to be one volt, if one joule of work is done in displacing a charge of one coulomb from infinity to that point, against the direction of electric field.*

- If V_A and V_B are the potentials of the points A and B respectively, then the potential difference between the points A and B is

$$V_{AB} = V_A - V_B = \frac{W_A - W_B}{q}$$

where   q = A small charge brought from infinity to the points A and B respectively.

W_A = Work done while bringing the charge q from infinity to the point A.

W_B = Work done while bringing the charge q from infinity to the point B.

$V_{AB} = V_A - V_B = 1$ volt if, $W_A - W_B = 1$ joule and $q = 1$ coulomb

3.3 POTENTIAL DUE TO A POINT CHARGE (Nov. 18)

- Consider a point charge +Q. 'A' is any point in the electric field of charge +Q.

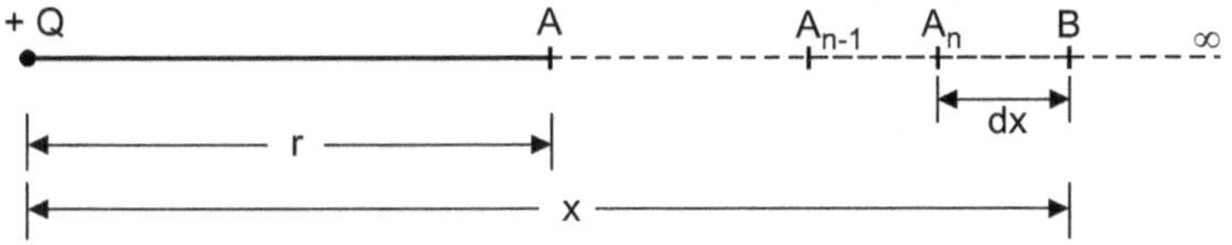

Fig. 3.6

Let,   r = Distance of 'A' from +Q

x = Distance of 'B' from +Q

ε_0 = Permittivity of free space

- Strength of the electric field at point 'A' due to +Q is,

$$E_A = \frac{1}{4\pi\varepsilon_0 k} \frac{Q}{r^2}$$

- Similarly, intensity at point 'B' due to +Q is,

$$E_B = \frac{1}{4\pi\varepsilon_0 k} \frac{Q}{x^2}$$

- In order to move a unit +ve charge by a very small distance dx against the field, some work is required.

- Let the work done be dw which is given by

$$\text{Work done} = \text{Force on unit +ve charge} \times \text{Displacement}$$

$$\text{Work done} = \text{Intensity of electric field} \times \text{Displacement at distance x}$$

$$\therefore \qquad dw = -\left(\frac{1}{4\pi\varepsilon_0 k} \frac{Q}{x^2}\right) dx$$

(–ve sign indicates that, work is done against the electric field)

- Thus, if unit positive charge is moved in very small steps from B to A, then total work done

$$W = dw_1 + dw_2 + dw_3 + \ldots$$

- This can be done by integration method.

$$\therefore \qquad V = \int dw$$

- We know that, potential at point 'A' is the total work done in bringing unit positive charge from ∞ (infinity) to point 'A'.

Putting the limits i.e. r and ∞,

$$\therefore \qquad V = \int_{\infty}^{r} dw$$

$$= \int_{\infty}^{r} -\frac{1}{4\pi\varepsilon_0 k} \frac{Q}{x^2} \, dx$$

$$= -\frac{Q}{4\pi\varepsilon_0 k} \int_{\infty}^{r} x^{-2} \, dx \qquad \text{but } \int x^{-2} = \frac{x^{-2+1}}{-2+1} = -\frac{1}{x}$$

$$= \frac{-Q}{4\pi\varepsilon_0 k} \left(\frac{-1}{x}\right)_{\infty}^{r}$$

$$= \frac{Q}{4\pi\varepsilon_0 kr} - 0$$

$$\therefore \qquad \boxed{V = \frac{Q}{4\pi\varepsilon_0 kr}} \quad \text{is the potential at 'A' due to +Q charge.}$$

$$\text{or} \qquad \boxed{V = 9 \times 10^9 \frac{Q}{kr}}$$

3.4 P.D. BETWEEN TWO POINTS

- Potential difference between two points is defined as the work done in carrying unit positive charge from one point to other point against the electric field.

- Consider charge +Q and point 'A' at a distance 'd' and point 'B' at a distance 'D' from charge +Q.

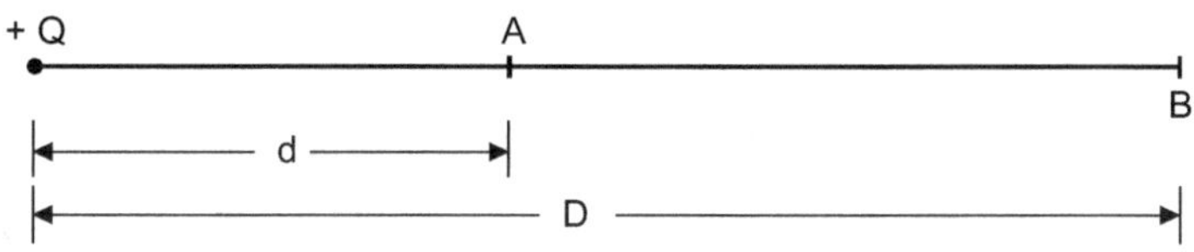

Fig. 3.7

- Now, potential difference between points 'A' and 'B' is defined as the work done in carrying unit positive charge from B to A, which is given by

$$V_{AB} = \frac{1}{4\pi\varepsilon_0} \frac{Q}{k} \left[\frac{1}{d} - \frac{1}{D}\right]$$

or

$$V_{AB} = 9 \times 10^9 \frac{Q}{k} \left[\frac{1}{d} - \frac{1}{D}\right]$$

Absolute Potential of a Point :

- The absolute potential of a point in an electric field is defined as *the amount of work done in carrying a unit positive charge from infinity (boundary of field) to that point.*

- Suppose the point is at a distance 'x' from charge Q, then the P.D. between this point and boundary is absolute potential at the point. The boundary of the electric field of the charge tends upto infinity.

Fig. 3.8

We have,
$$V_{AB} = 9 \times 10^9 \times \frac{Q}{k}\left[\frac{1}{d} - \frac{1}{D}\right]$$

Here $d = x$ and $D = \infty$

$\therefore$
$$V_{abs} = 9 \times 10^9 \times \frac{Q}{k}\left[\frac{1}{x} - \frac{1}{\infty}\right]$$

$\therefore$
$$V_{abs} = 9 \times 10^9 \times \frac{Q}{kx}$$
$\left(\because \ \frac{1}{\infty} = 0\right)$

or
$$V_{abs} = \frac{1}{4\pi\varepsilon_0} \frac{Q}{kx}$$

3.5 POTENTIAL OF A SPHERE OR POTENTIAL DUE TO A CHARGED SPHERE

- Suppose a charge +Q is placed on a sphere of radius r. Since all charges are similar, they try to push each other away. Therefore, all charges come to the surface and spread uniformly over the entire surface.

- Eventhough they are on the surface, their electric effects are as if they are all at the centre of the sphere.

- *Potential of a sphere is the amount of work done in carrying a unit positive charge from infinity (boundary of field) to the surface of the sphere.*

- Thus, potential of a sphere is the potential difference between a point on a surface and a point on the boundary of electric field.

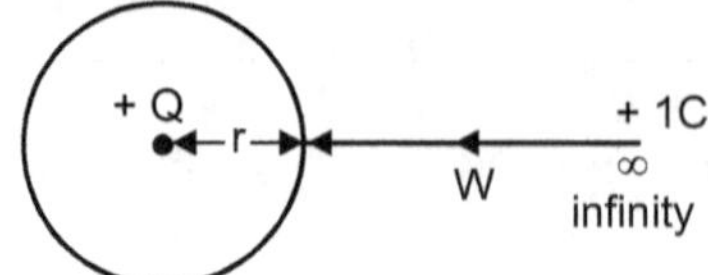

Fig. 3.9

- The distance of the point on the surface from the centre is 'r' and the distance of point on the boundary is ∞ (infinity).

i.e.
$$V = 9 \times 10^9 \cdot \frac{Q}{k}\left[\frac{1}{d} - \frac{1}{D}\right]$$

Here $d = r$ and $D = \infty$

∴
$$\boxed{V_{sph} = 9 \times 10^9 \cdot \frac{Q}{kr}}$$
$$\left(\because \ \frac{1}{\infty} = 0\right)$$

or
$$\boxed{V_{sph} = \frac{1}{4\pi\varepsilon_o} \cdot \frac{Q}{kr}}$$

Potential of Earth :

- Earth is fairly spherical body and its radius R is very large when compared to amount of charge and permittivity of the medium. Hence, we can approximately write $R = \infty$.

- Putting this value of 'R' in equation

$$V_{sph} = 9 \times 10^9 \cdot \frac{Q}{kr}$$

$$V_{earth} = 9 \times 10^9 \cdot \frac{Q}{kR}$$

But $R = \infty$

∴
$$V_{earth} = 9 \times 10^9 \cdot \frac{Q}{k \, \infty}$$
$$\left(\because \ \frac{1}{\infty} = 0\right)$$

∴
$$\boxed{V_{earth} = 0}$$

- Therefore earth is a body of zero potential. Hence the absolute potential of a point on the surface of the earth is always specified with respect to earth. For any amount of charge, radius of the earth remains infinitely large.

- Earthing is preferred in many electrical equipments which give way to leakage of current (charges), if any, to the earth, since charges flow from high potential to zero (low) potential.

3.6 DIELECTRIC STRENGTH

- It is the property of dielectrics. Dielectric material is a material which behaves as insulator (bad conductor of electricity). This dielectric behaves like insulator upto certain electric field intensity. If electric field intensity crosses certain limit, then it damages the dielectric and results into conduction. i.e. dielectric loses its insulating properties and becomes a conductor.

- *The magnitude of the electric field at which dielectric breakdown occurs in an insulating material is called dielectric strength of the material.*

- This dielectric strength depends on nature of material, time for which electric field is applied, temperature, humidity, magnitude of field.

Breakdown Potential :

- When electric field is applied to dielectric, then electric potential difference is created across it. If electric field increases, then potential difference across it increases. If we go on increasing the electric field and hence potential difference, then at certain stage dielectric loses its insulation and becomes a conductor. This limit is breakdown potential.

- *Breakdown potential is the potential difference which when applied across a unit thickness of the insulating medium damages the insulation.* It is measured in V/m.

SUMMARY

- Electric potential of a charge is the amount of work done in carrying a unit positive charge from infinity to the given charge.

- Electric potential (absolute potential) of a point in an electric field is the amount of work done in carrying a unit positive charge from infinity to that point.

- Electric P.D. between two points is defined as the amount of work done in carrying a unit positive charge from one point to the other point.

- Potential of a sphere is the amount of work done in carrying a unit positive charge from infinity to the surface of a charged sphere.

- Potential of earth is zero since it is a sphere of very large radius.

CONVERSIONS AND FORMULAE

1.　$1\ \mu C = 1 \times 10^{-6}\ C$

2.　$V_{AB} = 9 \times 10^9 \cdot \dfrac{Q}{k}\left[\dfrac{1}{d} - \dfrac{1}{D}\right]$

3.　$V = 9 \times 10^9 \cdot \dfrac{Q}{kx}$

4.　$V_{sph} = 9 \times 10^9\ \dfrac{Q}{kr}$

SOLVED EXAMPLES

Example 3.1 :

Calculate the potential at a point 20 cm away from the point charge +1C.

Solution :

Given :

$$d = 20\ cm = 0.2\ m,\ Q = +1C$$

$$V = 9 \times 10^9\ \frac{Q}{kd} \quad \text{Take } k = 1 \text{ assuming air medium}$$

$$= 9 \times 10^9\ \frac{(1)}{(1) \times (0.2)}$$

$$\boxed{V = 45 \times 10^9\ \text{volts}}$$

Example 3.2 :

Two charges of 33 μC and $-55\ \mu C$ are placed 1.6 m apart in air. Determine the position of the point in between them where the resultant potential is zero.

Solution :

$$Q_1 = 33\ \mu C \qquad\qquad Q_2 = -55\ \mu C$$

$$= 33 \times 10^{-6}\ C \qquad\qquad = -55 \times 10^{-6}\ C$$

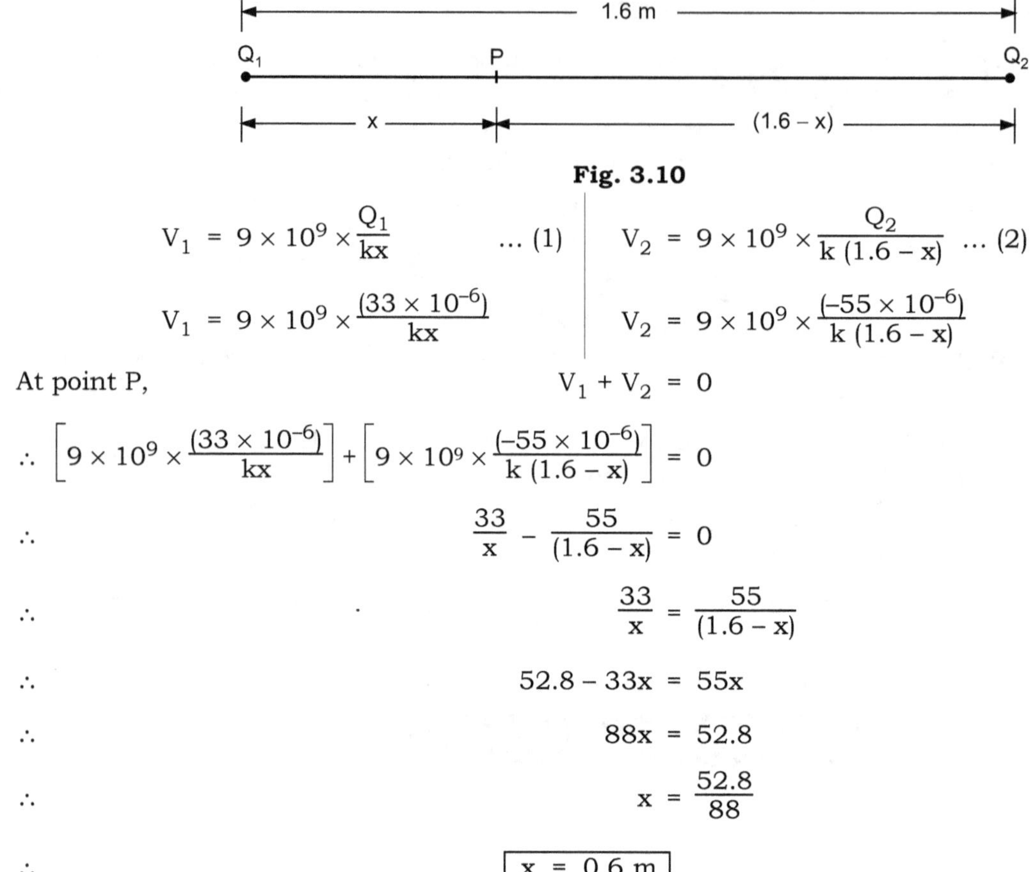

Fig. 3.10

$$V_1 = 9 \times 10^9 \times \frac{Q_1}{kx} \quad \dots (1) \qquad\qquad V_2 = 9 \times 10^9 \times \frac{Q_2}{k(1.6 - x)} \quad \dots (2)$$

$$V_1 = 9 \times 10^9 \times \frac{(33 \times 10^{-6})}{kx} \qquad\qquad V_2 = 9 \times 10^9 \times \frac{(-55 \times 10^{-6})}{k(1.6 - x)}$$

At point P, $\qquad\qquad\qquad\qquad\qquad V_1 + V_2 = 0$

$$\therefore \quad \left[9 \times 10^9 \times \frac{(33 \times 10^{-6})}{kx} \right] + \left[9 \times 10^9 \times \frac{(-55 \times 10^{-6})}{k(1.6 - x)} \right] = 0$$

$$\therefore \qquad\qquad\qquad \frac{33}{x} - \frac{55}{(1.6 - x)} = 0$$

$$\therefore \qquad\qquad\qquad \frac{33}{x} = \frac{55}{(1.6 - x)}$$

$$\therefore \qquad\qquad 52.8 - 33x = 55x$$

$$\therefore \qquad\qquad 88x = 52.8$$

$$\therefore \qquad\qquad x = \frac{52.8}{88}$$

$$\therefore \qquad\qquad \boxed{x = 0.6 \text{ m}}$$

Thus, the point P where resultant potential is zero is 0.6 m apart from 33 μC charge.

Example 3.3 :

An electric charge of 3×10^{-3} μC is placed at a point in the medium of dielectric constant 1.08. Find its potential at a point 20 cm away from it.

Solution :

Given : $Q = 3 \times 10^{-3}$ μC $= 3 \times 10^{-3} \times 10^{-6}$ C $= 3 \times 10^{-9}$ C.

$$k = 1.08, \quad d = 20 \text{ cm} = 0.2 \text{ m}$$

$$V = 9 \times 10^9 \times \frac{Q}{kd} = 9 \times 10^9 \times \frac{(3 \times 10^{-9})}{(1.08) \times (0.2)}$$

$$\therefore \qquad \boxed{V = 125 \text{ volts}}$$

Example 3.4 :

A sphere of diameter 16 cm is placed in air and given a charge of 0.0048 μC. What is the potential on its surface ?

Solution :

Given : $\qquad$ Diameter $= 16$ cm

$\therefore$ $\qquad\qquad\qquad$ r $=$ 8 cm $=$ 0.08 m

$\qquad\qquad\qquad$ Q $=$ 0.0048 μC $=$ 0.0048 $\times 10^{-6}$ C

$\qquad\qquad\qquad$ k $=$ 1 $\qquad\qquad$ ($\because$ air medium)

$\qquad\qquad$ V_{sph} $=$?

$\therefore$ $\qquad\qquad$ V_{sph} $=$ $9 \times 10^9 \times \dfrac{Q}{kr}$

$\qquad\qquad\qquad$ $=$ $9 \times 10^9 \times \dfrac{(0.0048 \times 10^{-6})}{(1) \times (0.08)}$

$$\boxed{V_{sph} \;=\; 540 \text{ V}}$$

Example 3.5 :

Two charges of 33 μC and 77 μC are placed 1.6 m apart in air. Determine the position of a point in between them where resultant potential is zero.

Solution :

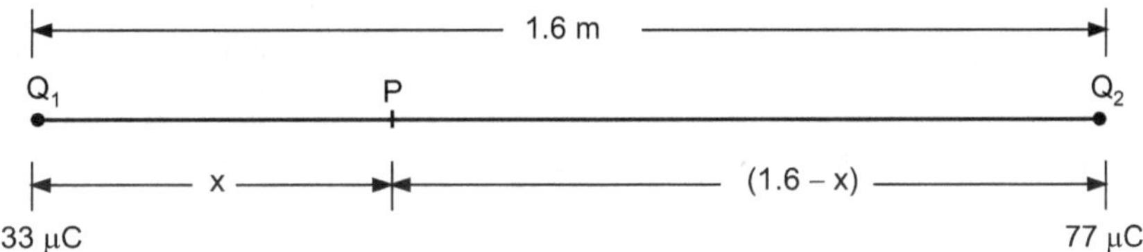

Fig. 3.11

Let 'P' be the point in between two charges where resultant potential is zero.

Let 'V_1' be the potential at 'P' due to Q_1. $\qquad$ Let 'V_2' be the potential at 'P' due to Q_2.

$V_1 = 9 \times 10^9 \times \dfrac{Q_1}{kx}$ $\qquad\qquad\qquad$ $V_2 = 9 \times 10^9 \times \dfrac{Q_2}{k(1.6-x)}$

$\qquad = 9 \times 10^9 \times \dfrac{(33 \times 10^{-6})}{(1) \times (x)}$ $\quad$ ... (1) $\qquad$ $= 9 \times 10^9 \times \dfrac{(77 \times 10^{-6})}{(1) \times (1.6-x)}$ $\quad$ ... (2)

At 'P', $\quad V_1 = V_2$

$9 \times 10^9 \times \dfrac{(33 \times 10^{-6})}{x} = 9 \times 10^9 \times \dfrac{(77 \times 10^{-6})}{(1.6-x)}$

$\dfrac{33}{x} = \dfrac{77}{(1.6-x)}$

$33 \times (1.6-x) = 77x$

$52.8 - 33x = 77x$

$52.8 = 110x$

$x = \dfrac{52.8}{110}$

$x = 0.48 \text{ m}$

EXERCISE

1. Define electric potential. State and define its S.I. units.

2. Derive the relation for potential due to a point charge at a given point in an electric field.

3. Derive expression for potential of a sphere.

4. Write a note on potential of earth.

5. Define absolute potential, potential difference.

6. Derive expression for potential difference between two points in an electric field of a charge.

7. Define absolute potential of a point and give its S.I. unit.

8. Define dielectric strength and breakdown potential.

9. Obtain an expression for electric potential at a point due to point charge using integration method.

10. Define potential of a point and potential difference between two points.

11. Why potential of the earth is said to be zero ?

PROBLEMS FOR PRACTICE

1. Calculate the potential due to a point charge of 1.6 μC at a point 40 cm away from the charge when placed in air.   (**Ans.** 36×10^3 V)

2. A sphere of diameter 16 cm is placed in a medium of dielectric constant 2.4 and is given a charge 0.0048 μC. Find (i) potential on its surface and (ii) potential at 6 cm away from its centre.

    (**Ans.** (i) 225 volts and (ii) 300 volts)

3. Two charges of 140 μC and –200 μC are placed 1 m apart in air. Find a point where the resultant potential is zero.   (**Ans.** 0.41 m from 140 μC)

4. Convert 4.8×10^{-20} eV into joules.   (**Ans.** 7.68×10^{-39} J)

    Hint : 1 eV = 1.6×10^{-19} J

4

CHAPTER

CAPACITY AND CONDENSERS

4.1   Introduction

4.2   Capacitance, Unit and Definition of 1 Farad

4.3   Principle of a Condenser (or a Capacitor)

4.4   Capacity of a Parallel-plate Condenser (Derivation)

4.5   Combination of Capacitances

Summary

Conversions

Important Formulae

Solved Examples

Exercise

Problems for Practice

4.1 INTRODUCTION

- We know that energy can be stored by compressing the spring, called potential energy. Similarly, energy from electric field can be stored in a device called capacitor, in the form of electrical potential energy. Hence, the capacitor has number of applications in electronics industry.

- If some charge is given to a conducting body, it is raised to a certain potential. If this charge is doubled, then the potential of a body is also doubled.

- Suppose a charge (Q_1) is given to a body and its potential is V_1. Now if charge is increased to $2Q_1$, its potential becomes $2V_1$. This means the ratio of charge to potential of a body remains constant. This constant is called capacity or capacitance of a conducting body.

4.2 CAPACITANCE, UNIT AND DEFINITION OF 1 FARAD

- **Capacitance (Definition) :** *The capacitance of a conductor (condenser) is defined as the ratio of charge on the conductor (condenser) to its potential.*

$$\text{Capacitance} = \frac{\text{Charge}}{\text{Potential}}$$

$$\therefore \qquad C = \frac{Q}{V}$$

If $\qquad V = 1$

then, $\qquad C = Q$

OR

- **Capacitance (Definition)** : *The capacitance of a conductor is also defined as the charge required to increase its potential by unity.*

- The unit of capacitance is farad.

$$1 \text{ farad } = \frac{1 \text{ coulomb}}{1 \text{ volt}}$$

- **One farad (Definition)** : *One farad of capacitance is defined as the capacitance of a conductor, the potential of which is increased by 1 volt by a charge of 1 coulomb.*

4.3 PRINCIPLE OF A CONDENSER (OR A CAPACITOR)

Condenser (capacitor) is used to store large amount of charge at very low potential. A condenser is an arrangement of two conductors to increase their capacitance.

1. Consider a metal plate A with positive charge Q on it. Let V_1 be its potential due to charge Q. Then $C_1 = \dfrac{Q}{V_1}$, where C_1 is the capacitance of a metal plate A.

   Thus to increase the capacitance, the potential V must be lowered.

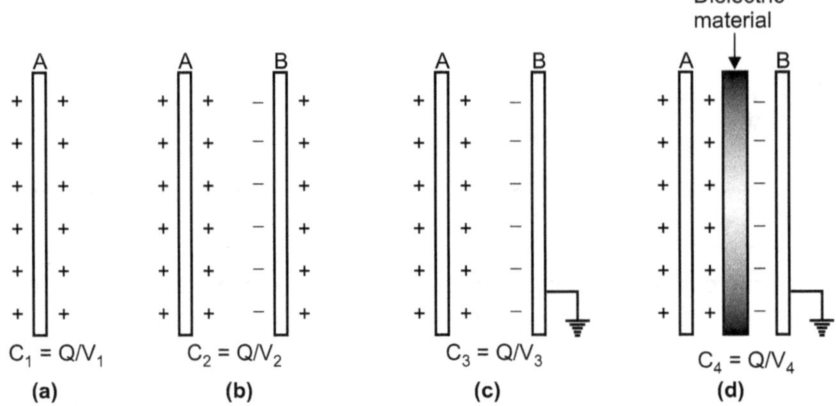

Fig. 4.1

2. In order to increase the capacitance of A by decreasing the potential of a metal plate A, another metal plate B is brought closer to A. The positive charge on A induces negative charge on inner side of B and positive charge on outer side. Thus, now positive charges on A are busy in holding negative charges on B. Therefore, potential of A decreases to V_2 and hence its capacitance increases to C_2.

$$\therefore \qquad C_2 = \frac{Q}{V_2}$$

3. Now metal plate B is connected to earth (zero potential). Positive charges on outer side of B which are free to move, go to earth; but negative charges on B being held by positive charges on A, do not leave the metal plate.

   As there are no positive charges on plate 'B', effective potential of plate A is further decreased to V_3 resulting into increase of capacitance to C_3.

$$\therefore \qquad C_3 = \frac{Q}{V_3}$$

4. Now, some dielectric material like mica sheet is introduced in between A and B, it is observed that capacitance of A again increases to C_4, this is because of the dielectric constant of medium inserted between them.

   Thus, in every stage capacitance of a metal plate A is increased and finally the arrangement shown in Fig. 4.1 (d) has more capacitance C_4. This arrangement is called *condenser*.

4.4 CAPACITY OF A PARALLEL-PLATE CONDENSER (DERIVATION)   (Nov. 18)

- A capacitor is an arrangement of two metal plates with a dielectric in between them. It is called a *parallel-plate condenser* (capacitor).

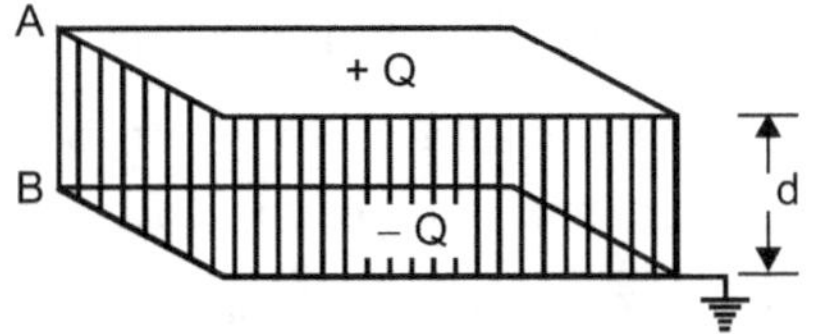

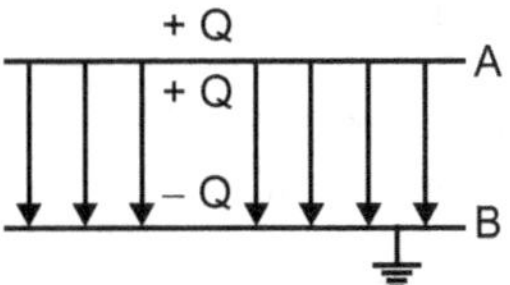

Fig. 4.2

- Consider two metal plates A and B.

  Let   A = Area of each metal plate.

         d = Distance between the two plates.

         k = Dielectric constant of the medium between them.

        +Q = Charge given to A.

        – Q = Charge induced on inner side of B.

         V = Potential difference between the two electrodes.

- Then, electric flux density D between the two plates is

$$D = \varepsilon_0 kE \qquad \qquad \dots (4.1)$$

   where   E = electric intensity

   and ε_0 = permittivity of free space

But $$D = \frac{\Psi}{A} = \frac{Q}{A} \qquad \qquad \dots (4.2)$$

   where ψ = electric flux

- Using equations (4.1) and (4.2), we can write

$$\frac{Q}{A} = \varepsilon_0 kE \qquad \qquad \text{But} \qquad \qquad E = \frac{V}{d}$$

$$\therefore \qquad \frac{Q}{A} = \varepsilon_0 k \left(\frac{V}{d}\right)$$

$$\therefore \qquad \frac{Q}{V} = \frac{\varepsilon_0 kA}{d} \qquad \qquad \text{But} \qquad \qquad C = \frac{Q}{V}$$

$$\therefore \qquad \boxed{C = \frac{\varepsilon_0 kA}{d}}$$

- This expression shows that capacitance 'C' can be increased by

1. increasing the area of metal plates,

2. increasing the dielectric constant k and

3. decreasing the distance between the two plates.

Factors affecting Capacitance of a Condenser :

The capacitance of a condenser depends upon following factors :

1. **Area of the plates :** If area of the plates of condenser increases, its capacitance increases. Hence capacitance of a condenser is directly proportional to the area of plates.

2. **The dielectric :** If a dielectric, like a plate of glass or ebonite, is inserted between the plates of a condenser, its potential decreases. Hence the capacitance of a condenser increases.

3. **The distance between the plates :** If the distance between the plates of a condenser is increased, its potential increases. Hence capacitance of a condenser decreases.

   Thus, the capacitance of a condenser is inversely proportional to the distance between the plates.

4.5 COMBINATION OF CAPACITANCES

4.5.1 Series Combination and Expression for Effective Capacitance (Condensers in Series)

- Consider three condensers of capacitances C_1, C_2 and C_3 connected in series between the points A and B.

- Let 'V' be the potential difference across the combination.

- Since the condensers are connected in series, total charge on each condenser is the same. But the potential V across the combination splits into three parts V_1, V_2 and V_3.

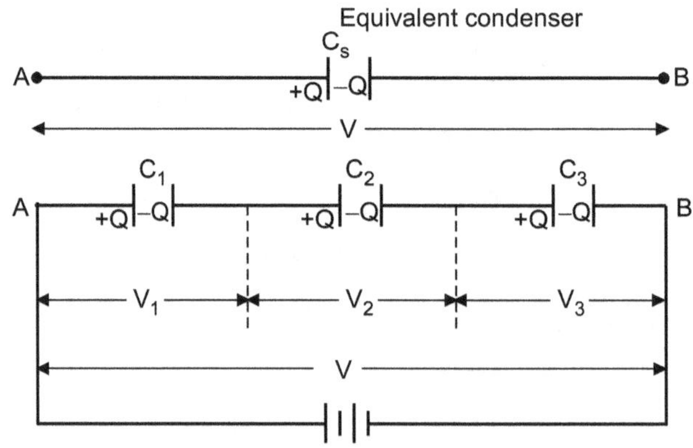

Fig. 4.3

- The values of V_1, V_2 and V_3 depend upon the values of C_1, C_2 and C_3.

$$\therefore \qquad V = V_1 + V_2 + V_3$$

But

$$C = \frac{Q}{V}$$

$$\therefore \qquad V = \frac{Q}{C}$$

$$\therefore \qquad V_1 = \frac{Q}{C_1}$$

$$V_2 = \frac{Q}{C_2}$$

$$V_3 = \frac{Q}{C_3}$$

and

$$V = \frac{Q}{C_s}$$

$$\therefore \qquad \frac{Q}{C_s} = \frac{Q}{C_1} + \frac{Q}{C_2} + \frac{Q}{C_3}$$

where C_s = equivalent capacitance of series combination

$$Q \times \frac{1}{C_s} = Q\left(\frac{1}{C_1} + \frac{1}{C_2} + \frac{1}{C_3}\right)$$

$$\therefore \qquad \frac{1}{C_s} = \frac{1}{C_1} + \frac{1}{C_2} + \frac{1}{C_3}$$

Law of condensers in series :

- **Statement :** Reciprocal of equivalent capacitance of the series combination is equal to the sum of reciprocals of capacitances of the condensers in series.

4.5.2 Parallel Combination and Expression for Effective Capacitance (Condensers in Parallel)

- Consider three condensers of capacitances C_1, C_2 and C_3 connected in between the points A and B.

- Let V be the potential difference across the combination.

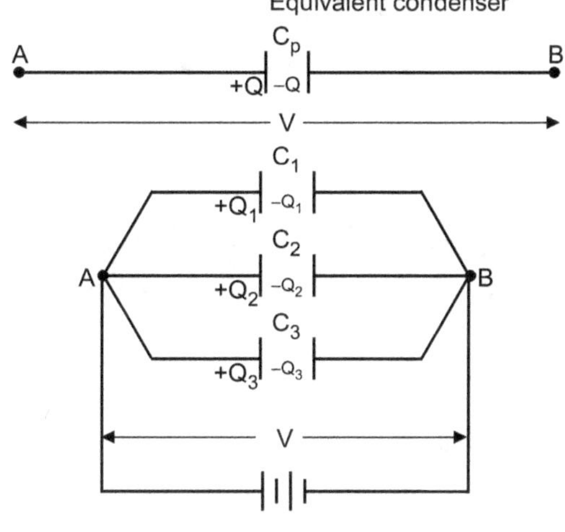

Fig. 4.4

- Since the condensers are connected in parallel, the potential across each condenser is the same.

- But the charge Q at point A splits into three parts say Q_1, Q_2 and Q_3.

- The distribution of Q_1, Q_2 and Q_3 depends upon the values of C_1, C_2 and C_3.

Thus
$$Q = Q_1 + Q_2 + Q_3$$

but
$$Q_1 = C_1 V$$
$$Q_2 = C_2 V$$
$$Q_3 = C_3 V$$

and
$$Q = C_p V$$

$\therefore$
$$C_p V = C_1 V + C_2 V + C_3 V$$

where C_p = equivalent capacitance of parallel combination

$\therefore$
$$C_p V = V (C_1 + C_2 + C_3)$$

$$\boxed{C_p = C_1 + C_2 + C_3}$$

Law of condensers in parallel :

- **Statement :** Equivalent capacitance of the parallel combination is equal to the sum of capacitances of the condensers in parallel.

SUMMARY

- Capacitance of a conductor is defined as the ratio of charge to its potential. Its unit is farad.

$$C = \frac{Q}{V}$$

- Condenser (Capacitor) is an arrangement made by keeping parallel plate near the charged plate in order to increase the capacitance of a given metal plate.

- Capacitance of a condenser increases with increase in area of plate, dielectric constant and by decreasing the distance between the two plates.

- If number of condensers are connected in series, then effective capacitance decreases and it becomes less than the least.

- If number of condensers are connected in parallel, then effective capacitance increases and it becomes equal to the sum of all capacitances.

CONVERSIONS

$$1 \ \mu F = 1 \times 10^{-6} \ F$$
$$1 \ nF = 1 \times 10^{-9} \ F$$
$$1 \ pF = 1 \times 10^{-12} \ F$$

IMPORTANT FORMULAE

1. $C = \dfrac{Q}{V}$

2. $C = \dfrac{\varepsilon_o \, kA}{d}$

3. $\dfrac{1}{C_s} = \dfrac{1}{C_1} + \dfrac{1}{C_2} + \dfrac{1}{C_3}$

4. $C_p = C_1 + C_2 + C_3$

where,

C = Capacity of a condenser in farad

k = Dielectric constant

A = Area of plate

d = Distance between two plates

Q = Charge

V = P.D. between two plates

C_s = Equivalent capacitance in series combination

C_p = Equivalent capacitance in parallel combination

SOLVED EXAMPLES

Example 4.1 :

The potential difference of 60 volts is applied across a capacitor of capacitance 20 µF. Calculate the charge on the plates.

Solution :

Given : $V = 60$ V, $C = 20$ µF $= (20 \times 10^{-6})$ F, $Q = ?$

We have, $C = \dfrac{Q}{V}$

$\therefore$ $Q = C \times V$

$= (20 \times 10^{-6}) \times (60)$

$\boxed{Q = 1200 \times 10^{-6} \text{ C}}$ or $\boxed{Q = 1200 \text{ µC}}$

Example 4.2 :

A charge of 2 µC raises the potential of a capacitor by 50 V. Calculate its capacitance.

Solution :

Given : $Q = 2$ µC $= 2 \times 10^{-6}$ C, $V = 50$ V, $C = ?$

We have, $C = \dfrac{Q}{V}$

$= \dfrac{2 \times 10^{-6}}{50}$

$\boxed{C = 0.04 \times 10^{-6} \text{ F}}$ or $\boxed{C = 0.04 \text{ µF}}$

Example 4.3 :

Calculate the potential difference across a capacitor of capacitance 500 µF and if charge on plates is 1500 µC.

Solution :

Given : $C = 500$ µF $= (500 \times 10^{-6})$ F, $Q = 1500$ µC $= (1500 \times 10^{-6})$ C, $V = ?$

We have, $C = \dfrac{Q}{V}$

$\therefore$ $V = \dfrac{Q}{C}$

$= \dfrac{(1500 \times 10^{-6})}{(500 \times 10^{-6})}$

$\boxed{V = 3 \text{ V}}$

Example 4.4 :

Area of a parallel-plate condenser is 3.21 m^2 and distance between the plates is 0.1 mm. Find the capacitance of a condenser if its dielectric constant is 7 and $\varepsilon_o = 8.9 \times 10^{-12}$.

Solution :

Given : $A = 3.21$ m^2, $d = 0.1$ mm $= 0.1 \times 10^{-3}$ m, $C = ?$

$k = 7$, $\varepsilon_o = 8.9 \times 10^{-12}$

$$C = \frac{\varepsilon_0 kA}{d}$$

$$= \frac{(8.9 \times 10^{-12}) \times (7) \times (3.21)}{(0.1 \times 10^{-3})}$$

$\therefore$
$$\boxed{C = 1.99 \times 10^{-6} \text{ F}}$$

or
$$\boxed{C = 1.99 \ \mu\text{F}}$$

Example 4.5 :

Area of a parallel-plate condenser is 2 m² and its capacitance is 2 μF. Find the distance between two plates of a condenser if the dielectric constant is 4 and $\varepsilon_0 = 8.9 \times 10^{-12}$.

Solution :

Given :

$A = 2 \text{ m}^2$

$C = 2 \ \mu\text{F} = 2 \times 10^{-6} \text{ F}$

$d = ?$

$k = 4$

$\varepsilon_0 = 8.9 \times 10^{-12}$

We have,
$$C = \frac{\varepsilon_0 kA}{d}$$

$\therefore$
$$d = \frac{\varepsilon_0 kA}{C}$$

$$= \frac{(8.9 \times 10^{-12} \times 4 \times 2)}{(2 \times 10^{-6})}$$

$$\boxed{d = 3.56 \times 10^{-5} \text{ m}}$$

Example 4.6 :

Find the area of parallel-plate condenser if its capacitance is 3 μF, distance between the two plates is 0.04 mm, dielectric constant is 6 and $\varepsilon_0 = 8.9 \times 10^{-12}$.

Solution :

Given :

$C = 3 \ \mu\text{F} = 3 \times 10^{-6} \text{ F}$

$d = 0.04 \text{ mm} = 0.04 \times 10^{-3} \text{ m}$

$k = 6$

$\varepsilon_0 = 8.9 \times 10^{-12}$

$A = ?$

We have
$$C = \frac{\varepsilon_0 kA}{d}$$

$\therefore$
$$\frac{C \times d}{\varepsilon_0 k} = A$$

$\therefore$
$$A = \frac{3 \times 10^{-6} \times 0.04 \times 10^{-3}}{(8.9 \times 10^{-12} \times 6)}$$

$$\boxed{A = 2.247 \text{ m}^2}$$

Example 4.7 :

Three condensers of capacitances 6, 12 and 16 μF are connected in series. A potential difference of 220 volt is applied to the combination. How much charge will be drawn ?

Solution :

Given : Let $C_1 = 6$ μF, $C_2 = 12$ μF and $C_3 = 16$ μF

For series combination, $\dfrac{1}{C_s} = \dfrac{1}{C_1} + \dfrac{1}{C_2} + \dfrac{1}{C_3}$

$\therefore \qquad \dfrac{1}{C_s} = \dfrac{1}{6} + \dfrac{1}{12} + \dfrac{1}{16}$

$\therefore \qquad \dfrac{1}{C_s} = \dfrac{8 + 4 + 3}{48}$

$\therefore \qquad \dfrac{1}{C_s} = \dfrac{15}{48}$

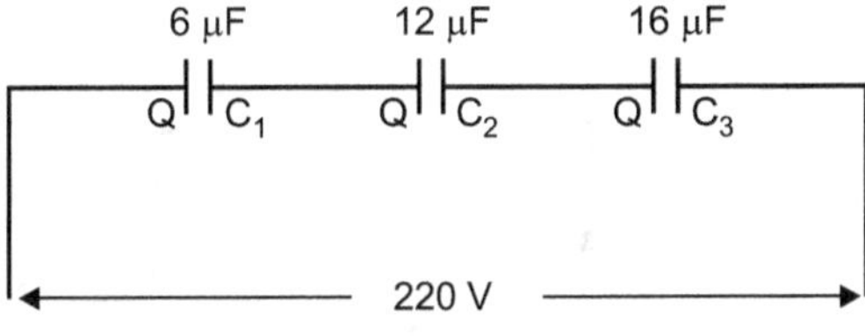

Fig. 4.5

$\therefore \qquad C_s = \dfrac{48}{15}$

$\therefore \qquad C_s = 3.2$ μF

Now, $\qquad C = \dfrac{Q}{V}$

$\therefore \qquad Q = CV = (3.2 \times 10^{-6}) \times (220)$

$\qquad Q = 704 \times 10^{-6}$ C

or $\qquad \boxed{Q = 704 \text{ μC}}$

Example 4.8 :

Two condensers have an equivalent capacitance of 12 μF when connected in parallel and 2.25 μF when connected in series. Calculate their individual capacitances.

Solution :

Given : $\quad C_p = 12$ μF, $C_s = 2.25$ μF, $C_1 = ?$, $C_2 = ?$

For parallel combination,

$$C_p = C_1 + C_2$$

$$12 = C_1 + C_2 \qquad \dots (1)$$

For series combination,

$$\dfrac{1}{C_s} = \dfrac{1}{C_1} + \dfrac{1}{C_2}$$

$$\dfrac{1}{C_s} = \dfrac{C_2 + C_1}{C_1 C_2}$$

$$\therefore \qquad C_s = \dfrac{C_1 C_2}{C_2 + C_1}$$

$$2.25 = \dfrac{C_1 C_2}{C_2 + C_1} \qquad \dots (2)$$

From equation (1), $C_1 = 12 - C_2$. Putting this value in equation (2), we get

$$2.25 = \frac{(12 - C_2)\, C_2}{C_2 + (12 - C_2)}$$

$$2.25 = \frac{12\, C_2 - C_2^2}{12}$$

$$27.00 = 12\, C_2 - C_2^2$$

$$C_2^2 - 12\, C_2 + 27 = 0$$

$$(C_2 - 9)\,(C_2 - 3) = 0$$

$\therefore$

| $C_2 = 9\ \mu F$ and $C_1 = 3\ \mu F$ | or | $C_2 = 3\ \mu F$ and $C_1 = 9\ \mu F$ |

Example 4.9 :

Find the resultant capacity of the following combination.

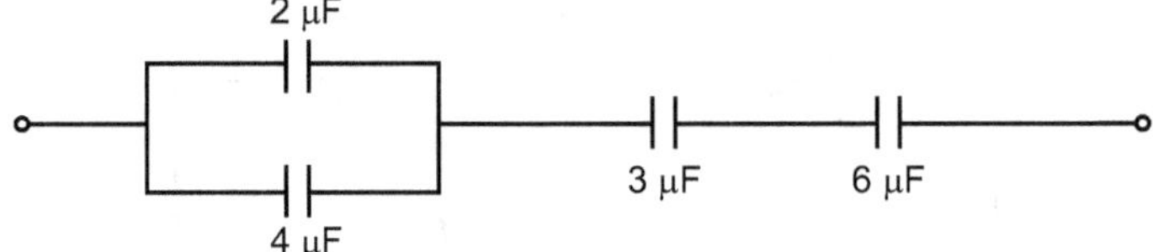

Fig. 4.6

Solution : $C_p = 2 + 4 = 6\ \mu F$

Now consider

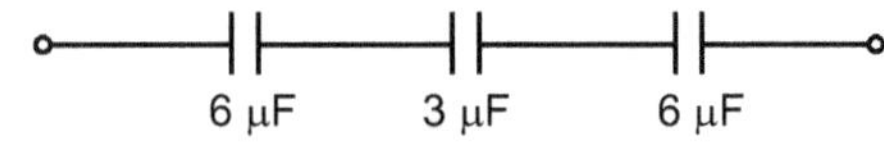

Fig. 4.7

$$\frac{1}{C_s} = \frac{1}{6} + \frac{1}{3} + \frac{1}{6}$$

$\therefore$
$$\frac{1}{C_s} = \frac{1 + 2 + 1}{6} = \frac{4}{6}$$

$\therefore$
$$C_s = \frac{6}{4} = \frac{3}{2}$$

$$\boxed{C_s = 1.5\ \mu F}$$

Example 4.10 :

Two condensers of capacitances 15 µF and 10 µF are connected in parallel across a battery of 12 volt. Find the resultant capacitance and charge on each condenser.

Solution :

Given :

$$C_1 = 15\ \mu F$$
$$= 15 \times 10^{-6}\ F$$
$$C_2 = 10\ \mu F$$
$$= 10 \times 10^{-6}\ F$$
$$V = 12\ V$$

Resultant capacitance, $C_p = C_1 + C_2$

$$= 15 + 10$$

$$\boxed{C_p = 25\ \mu F}$$

Fig. 4.8

$$\text{Voltage across } C_1 = \text{Voltage across } C_2 = V \qquad (\because \text{ it is a parallel combination})$$

$$Q_1 = C_1 V$$
$$= (15 \times 10^{-6})(12)$$

$$\boxed{Q_1 = 180 \times 10^{-6} \text{ C}} \quad \text{or} \quad \boxed{Q_1 = 180 \ \mu C}$$

$$Q_2 = C_2 V$$
$$= (10 \times 10^{-6})(12)$$

$$\boxed{Q_2 = 120 \times 10^{-6} \text{ C}} \quad \text{or} \quad \boxed{Q_2 = 120 \ \mu C}$$

Example 4.11 :

Find the resultant capacity between points A and B of the following combination.

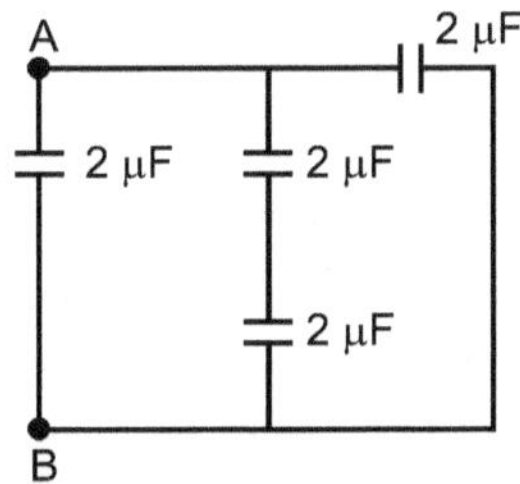

Fig. 4.9

Solution :

Step 1 : Above circuit diagram can be drawn as follows :

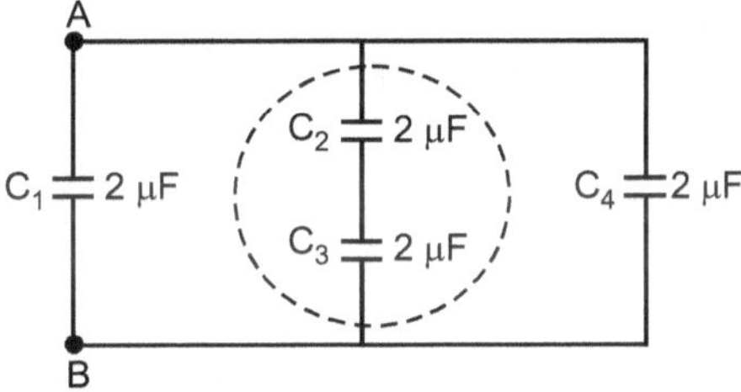

Fig. 4.10

C_2 and C_3 are in series.

$$\therefore \qquad \frac{1}{C_S} = \frac{1}{C_2} + \frac{1}{C_3}$$

$$\therefore \qquad \frac{1}{C_S} = \frac{1}{2} + \frac{1}{2}$$

$$\therefore \qquad \boxed{C_S = 1 \ \mu F}$$

Step 2 :

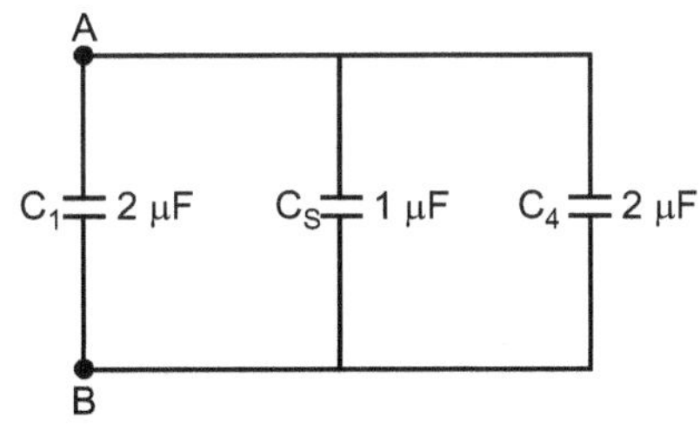

Fig. 4.11

C_1, C_S and C_4 are in parallel.

$$\therefore \quad C_p = C_1 + C_S + C_4$$

$$= 2 + 1 + 2$$

$$\therefore \quad \boxed{C_p = 5 \ \mu F}$$

Example 4.12 :

Area of a parallel-plate condenser is 1.6 m² and distance between the two plates is 1.2 mm. Dielectric constant of the material between the two plates is 3. Find capacitance of the condenser.

Solution :

Given :

$$A = 1.6 \ m^2$$

$$d = 1.2 \ mm = 1.2 \times 10^{-3} \ m$$

$$k = 3$$

$$C = ?$$

$$V = 240 \ volts$$

$$E = ?$$

$$\varepsilon_0 = 8.85 \times 10^{-12} \ C^2/Nm^2$$

We have,

$$C = \frac{\varepsilon_0 k A}{d}$$

$$C = \frac{(8.85 \times 10^{-12}) \times (3) \times (1.6)}{(1.2 \times 10^{-3})}$$

$$\boxed{C = 3.54 \times 10^{-8} \ F}$$

Example 4.13 :

Two capacitors of 1 μF and 2 μF are connected in series across a 60 V d.c. supply. Calculate (a) equivalent capacitance, (b) charge on each condenser, (c) potential drop across each condenser.

Solution :

Given : $C_1 = 1 \ \mu F = 1 \times 10^{-6} \ F$, $\ C_2 = 2 \ \mu F = 2 \times 10^{-6} \ F$, $\ V = 60 \ V$

In series : (a) Equivalent capacitance :

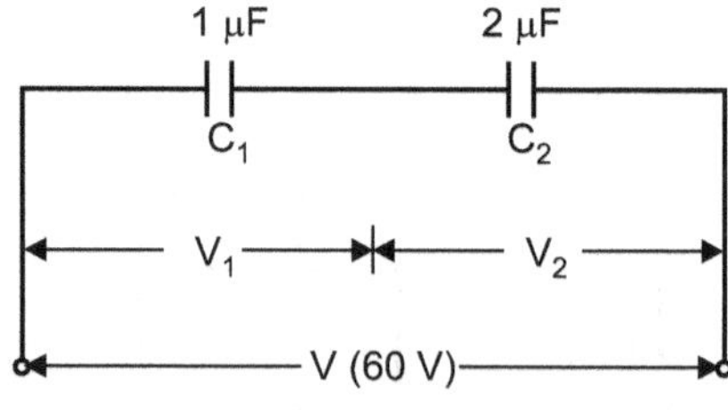

Fig. 4.12

(a) For series combination equivalent capacitance is given by

$$\frac{1}{C_S} = \frac{1}{C_1} + \frac{1}{C_2}$$

$$\frac{1}{C_S} = \frac{1}{1} + \frac{1}{2}$$

$$\frac{1}{C_S} = \frac{(2 + 1)}{2}$$

$\therefore \qquad C_S = \dfrac{2}{3}$

$\therefore \qquad \boxed{C_S = 0.667 \ \mu F}$

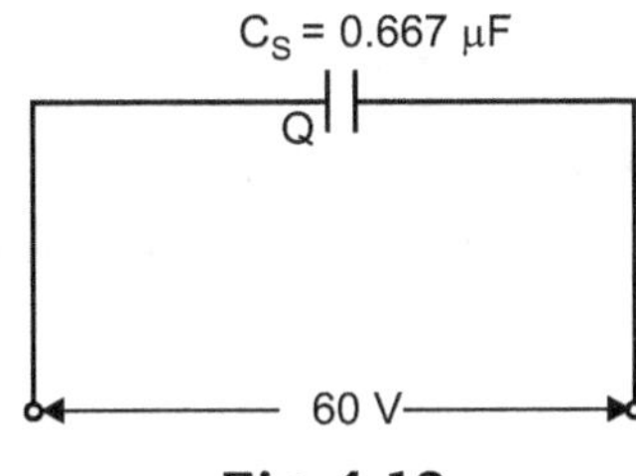

Fig. 4.13

(b) Charge on each condenser :

We have, $\qquad\qquad C_S = \dfrac{Q}{V}$

$\therefore \qquad\qquad Q = C_S \times V = (0.667 \times 10^{-6}) \times 60$

$\boxed{Q = 40 \times 10^{-6} \ C} \quad$ or $\quad \boxed{Q = 40 \ \mu C}$

In series, the charge on each condenser is **same** (common).

Thus,

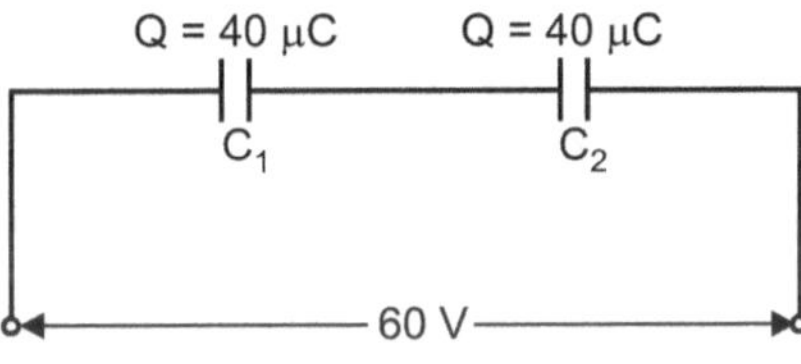

Fig. 4.14

(c) Potential drop across each condenser : We have

$$C_1 = \frac{Q}{V_1} \qquad\qquad \text{and} \qquad\qquad C_2 = \frac{Q}{V_2}$$

$\therefore \qquad V_1 = \dfrac{Q}{C_1} \qquad\qquad\qquad\qquad V_2 = \dfrac{Q}{C_2}$

$$= \frac{(40 \times 10^{-6})}{(1 \times 10^{-6})} \qquad\qquad\qquad = \frac{(40 \times 10^{-6})}{(2 \times 10^{-6})}$$

$\boxed{V_1 = 40 \ V} \qquad\qquad\qquad\qquad \boxed{V_2 = 20 \ V}$

Example 4.14 :

Two capacitors of 1 μF and 2 μF are connected in parallel across a 60 V d.c. supply. Calculate (a) equivalent capacitance, (b) potential drop across each condenser, (c) charge on each condenser.

Solution :

In parallel :

(a) Equivalent capacitance :

$$C_p = C_1 + C_2 = 1 + 2$$

$$\boxed{C_p = 3 \ \mu F}$$

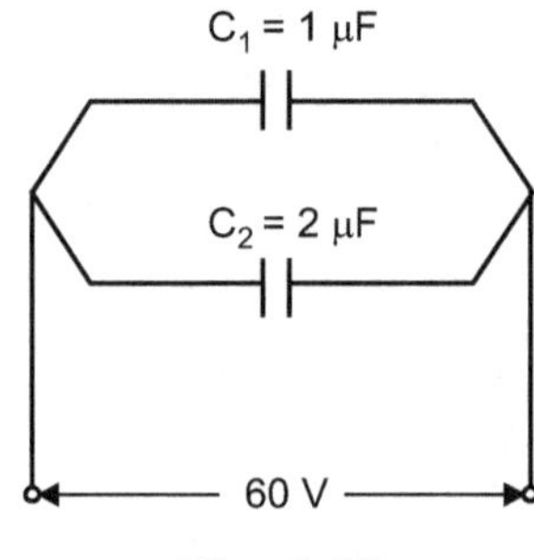

Fig. 4.15

(b) Potential drop across each condenser : In parallel combination, potential drop across each condenser and also across equivalent condenser is common i.e. 60 V.

$\therefore$ $\boxed{\text{P.D. across each condenser} = 60\ \text{V}}$

(c) Charge on each condenser :

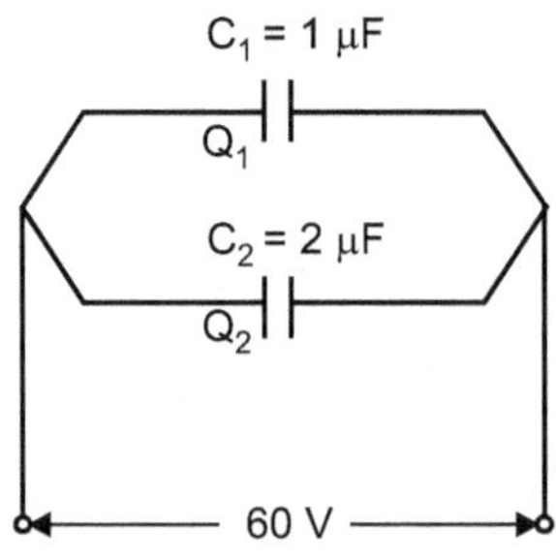

Fig. 4.16

We have $C = \dfrac{Q}{V}$

$\therefore$ $C_1 = \dfrac{Q_1}{V}$ $C_2 = \dfrac{Q_2}{V}$

$\therefore$ $Q_1 = C_1 \times V$ $Q_2 = C_2 \times V$

       $= (1 \times 10^{-6}) \times (60)$ $= (2 \times 10^{-6}) \times 60$

$\boxed{Q_1 = 60 \times 10^{-6}\ \text{C}}$ $\boxed{Q_2 = 120 \times 10^{-6}\ \text{C}}$

or   or

$\boxed{Q_1 = 60\ \mu C}$ $\boxed{Q_2 = 120\ \mu C}$

EXERCISE

1. Define capacity of a conductor.

2. State formula for capacitance of a parallel-plate capacitor.

3. State formula for the resultant capacitance when condensers are connected in series.

4. State formula for the resultant capacitance when condensers are connected in parallel.

5. Define capacity of a condenser.

6. State formula for capacity of a parallel-plate condenser and state factors affecting capacitance of a condenser and give their relation.

7. Explain principle of condenser.

8. Define unit capacitor or 1 farad capacity.

9. What is capacitance ? State its S.I. unit.

10. Define condenser and 1 farad.

11. Explain principle of capacitor.

12. Define one farad capacity of condenser.

13. Obtain an expression for capacitance of a parallel-plate condenser.

14. Show that $C = \dfrac{\varepsilon_0 kA}{d}$ in case of parallel-plate condenser, where symbols have usual meanings.

15. Obtain an expression for equivalent capacitance of capacitors connected in parallel combination.

16. Obtain the expression for equivalent capacitance of 'n' capacitors connected in (a) series and (b) parallel.

17. State and prove parallel law of condensers.

18. Derive an expression for effective capacitance when three capacitors are connected in series with each other.

19. Draw circuit diagrams and symbols of (i) condensers in parallel, (ii) condensers in series.

20. State expression for capacitance of a parallel-plate condenser filled with dielectric.

PROBLEMS FOR PRACTICE

1. The potential difference between the plates of 0.5 μF capacitor is 200 V. Calculate the charge on the plate. **(Ans.** $Q = 100 \times 10^{-6}$ C)

2. When a charge of 0.05 μC is given to a conductor, its potential is raised to 100 V. Find its capacitance. **(Ans.** $C = 5 \times 10^{-10}$ F)

3. Show how will you connect two capacitances each of 4 μF so that the resultant capacitance is 2 μF ? **(Ans.** In series (also draw diagram)

4. Two capacitors 12 μF and 24 μF are connected in series. Find equivalent capacitance.

   (Ans. $C_S = 8$ μF)

5. Two condensers when connected in series and parallel have equivalent capacitances of 3 farad and 16 farad respectively. What is the capacitance of each condenser ?

   (Ans. $C_1 = 12$ farad and $C_2 = 4$ farad)

6. Three condensers are connected in series across 75 V supply. The voltages across them are 20, 25, 30 V respectively and the charge on each is 3×10^{-3} C. Find the capacity of each condenser and also the combination.

   (Ans. $C_1 = 0.15 \times 10^{-3}$ F ; $C_2 = 0.12 \times 10^{-3}$ F

   $C_3 = 0.1 \times 10^{-3}$ F and $C_s = 0.04 \times 10^{-3}$ F)

7. Calculate the potential difference across a capacitor of capacitance 1200 μF and if charge on the plates is 800 μC. **(Ans.** V = 0.67 V)

8. Area of parallel-plate condenser is 3 m² and its capacitance is 3 μF. Find the distance between the two plates of condenser if the dielectric constant is 4 and $\varepsilon_0 = 8.9 \times 10^{-12}$.

(**Ans.** $d = 3.56 \times 10^{-5}$ m)

9. Area of parallel-plate condenser is 0.7 m² and distance between the two plates is 2 mm. Dielectric constant of the material between the two plates is 5. Find capacitance of the condenser. (**Ans.** $C = 1.549 \times 10^{-8}$ F)

10. Find the equivalent capacitance of the following combination. (**Ans.** $C_{eq.} = 1$ μF)

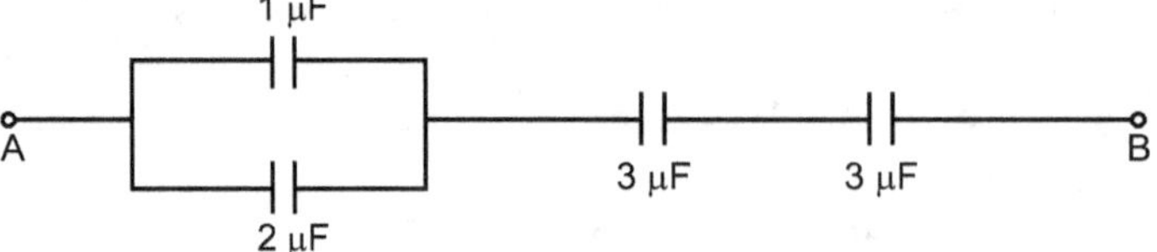

Fig. 4.17

11. Find the resultant capacitance between A and B of the following combination.

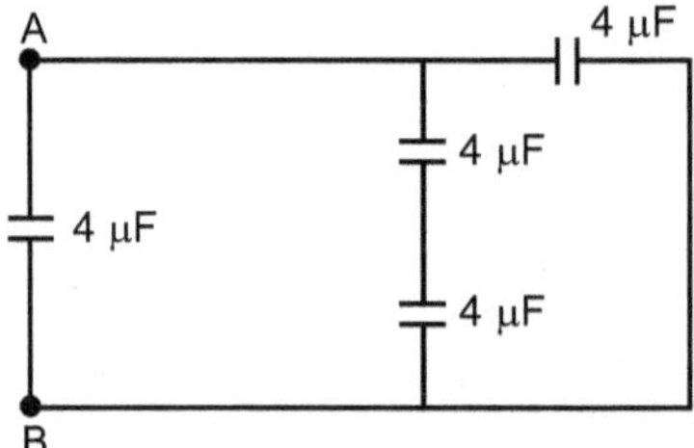

Fig. 4.18

(**Ans.** $C_{eq} = 10$ μF)

12. Two capacitors of 2 μF and 4 μF are connected in series across 30 V d.c. supply. Calculate : (a) equivalent capacitance, (b) charge on each condenser, (c) potential drop across each condenser. (**Ans.** (a) $C_S = 1.33$ μF, (b) $Q = 39.9 \times 10^{-6}$ C, (c) $V_1 = 20$ V and $V_2 = 10$ V)

13. Two capacitors of 2 μF and 4 μF are connected in parallel across 30 V d.c. supply. Calculate (a) equivalent capacitance, (b) potential drop across each condenser, (c) charge on each condenser. (**Ans.** (a) $C_p = 6$ μF, (b) P.D. $\rightarrow$ V = 30 V, (c) $Q_1 = 60$ μC and $Q_2 = 120$ μC)

5

CHAPTER

CURRENT ELECTRICITY

5.1   Introduction

5.2   Electric Current (I)

5.3   Resistance (R)

5.4   Resistivity or Specific Resistance (ρ)

5.5   Conductance (G)

5.6   Conductivity or Specific Conductance (σ)

5.7   Ohm's Law

5.8   Combination of Resistances

5.9   Kirchhoff's Laws

5.10 Application of Kirchhoff's Laws to Wheatstone's Bridge Circuits

Summary

Important Formulae

Solved Examples

Exercise

Problems for Practice

5.1 INTRODUCTION

- We know that almost all metals are good conductors of electricity and wood, plastic, rubber are the examples of bad conducting materials of electricity.

- **Atomic structure :** An atom consists of a centrally placed heavy nucleus which consists of **proton** carrying positive charge and **neutron**, which is electrically neutral.

- An electron carrying negative charge revolves around the nucleus. The electrostatic force of attraction between proton (positive) and electron (negative) keeps the electron in circular motion.

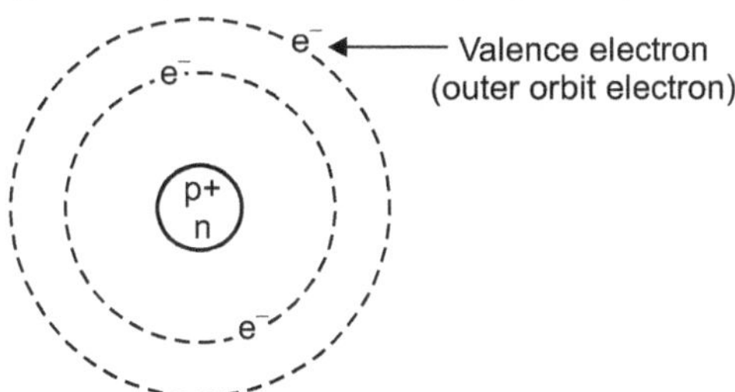

Fig. 5.1

- In case of metals, the valence electrons (outer) are loosely held by nucleus.

- Hence these electrons can be easily displaced by small external electrical energy (force).
- The external electrical energy required is called electromotive force (e.m.f.).
- As water flows from higher level to lower level through a pipeline, electric current flows from high electric potential to low electric potential. When electric potential difference (P.D.) is applied to a conductor i.e. if battery is connected across a conductor, then charges start flowing in a particular direction.

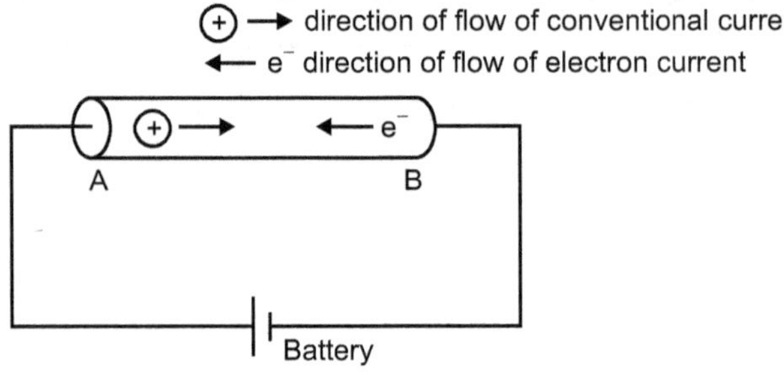

Fig. 5.2

Electron current :

- It is the current due to flow of electrons.
- This current flows externally from negative of battery supply to positive of battery i.e. from end B to end A of the conductor as shown.

Conventional current :

- The direction of flow of conventional current is opposite to the electron flow. Thus, conventional current flows from end A to end B through the conductor as shown. It is the direction of flow of positive charges.

5.2 ELECTRIC CURRENT (I)

- **Electric current (Definition) :** It is defined as the rate of flow of electric charge through cross-section of conducting wire in the electrical circuit.
- **Definition :** Electric current (I) is defined as the flow of electric charge per unit time.

OR

- **Definition :** The amount of electric charge flowing (through a circuit) in one second is called as electric current.

$$\text{Current} = \frac{\text{Charge}}{\text{Time}}$$

$$I = \frac{Q}{t}$$

- **S.I.** unit of **current** is **amperes** i.e. **A**.

$$I = \frac{Q}{t} \qquad \text{If} \quad Q = 1\ C$$

$$t = 1\ s$$

$$\text{then}\ I = 1\ A$$

One ampere (1 A) :

- **Definition :** A current of 1 A is said to be flowing through a circuit when a charge of 1 C crosses a point in time 1 sec.

OR

- **Definition :** When a charge of 1 coulomb (i.e. 6.25×10^{18} electrons) is passing through a point in 1 second, then current of 1 A is said to be flowing through a circuit.

5.3 RESISTANCE (R)

- **Definition :** The resistance of a material of wire is defined as the opposition offered by a material to the flow of current.

OR

- **Definition :** The resistance is the property of the material by virtue of which it opposes the flow of current through it.

- Low resistance means good conductor of electricity. e.g. copper, iron etc.

- High resistance means bad conductor of electricity. e.g. wood, plastic, rubber.

- **S.I.** unit of **resistance** is **ohm**, it is shown by a symbol Ω.

$$R = \frac{V}{I}$$

$$1 \text{ ohm} = \frac{1 \text{ volt}}{1 \text{ ampere}}$$

- **1 ohm (Definition) :** If a potential difference of 1 volt is applied across a conductor and it produces a current of 1 ampere through it, then the resistance of a conductor is said to be 1 Ω.

5.4 RESISTIVITY OR SPECIFIC RESISTANCE (ρ) (RHO)

- **Factors affecting resistance of a conductor :**

   (1) Length of a conductor.

   (2) Cross-sectional area of a conductor.

   (3) Temperature of a conductor.

   (4) Material of conductor.

- Practically, it is observed that, resistance of a given material of a wire at a constant temperature is directly proportional to its length and inversely proportional to its cross-sectional area.

$$R \propto \frac{L}{A} \qquad\qquad \text{where, } R = \text{Resistance of a wire}$$

$$\therefore \quad R = \text{Constant} \times \frac{L}{A} \qquad\qquad L = \text{Length of a wire}$$

$$R = \rho \times \frac{L}{A} \qquad\qquad A = \text{Cross-sectional area}$$

$$\therefore \quad \rho = \frac{R \times A}{L} \; \frac{\Omega m^2}{m} \; \dots \; \Omega m \qquad\qquad \rho = \text{Specific resistance or resistivity}$$

which is constant for a given material.

- **Unit of resistivity or S.I. unit of specific resistance is ohm-meter i.e. Ωm**. Other unit of ρ is ohm-centimeter i.e. Ω cm.

We have, $\rho = \dfrac{RA}{L}$

If A = 1 cm², L = 1 cm then $\rho = R$

- **Specific resistance or Resistivity (ρ) (Definition) :** Specific resistance of a given material of wire is defined as a resistance of a wire of unit length and unit cross-sectional area.

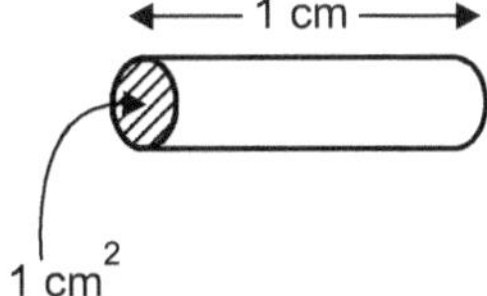

Resistance R of such a metal piece is the specific resistance of that metal.

Fig. 5.3

5.5 CONDUCTANCE (G)

- **Definition :** Conductance is defined as the reciprocal of resistance.

$$G = \frac{1}{R}$$

- **Unit** of **conductance** is $/\Omega$. S.I. unit of conductance is **siemens – S** or mho (Ω^{-1}) or $(\mho)$.

5.6 CONDUCTIVITY OR SPECIFIC CONDUCTANCE (σ) (SIGMA)

- **Definition :** It is defined as the reciprocal of resistivity.

OR

- **Definition :** It is defined as the reciprocal of specific resistance.

$$\therefore \text{ Conductivity (specific conductance)} = \frac{1}{\text{Resistivity}} \text{ or } \frac{1}{\text{Specific resistance}}$$

$$\sigma = \frac{1}{\rho}$$

- **Unit of conductivity** is $/\Omega$m. S.I. unit of conductivity is **siemens/meter** … **S/m**.

5.7 OHM'S LAW

- **Ohm's law - Statement :** Physical state of a conductor (material, length, area and temperature) remaining the same, the electric current flowing through a conductor is directly proportional to the potential difference across it.

$$I \propto V$$

i.e. $\quad V \propto I$

$$V = \text{Constant} \times I$$

$$\frac{V}{I} = \text{Constant} = R$$

OR

- **Ohm's law - Statement :** When the physical state of a metallic conductor remains constant, the electric current through it is directly proportional to the potential difference between the ends of a conductor.

- The physical state means length, area, temperature, etc.

OR

- **Ohm's law - Statement :** Physical state of a conductor (material, length, area and temperature) remaining the same, the ratio of potential difference (V) to current (I) is always constant.

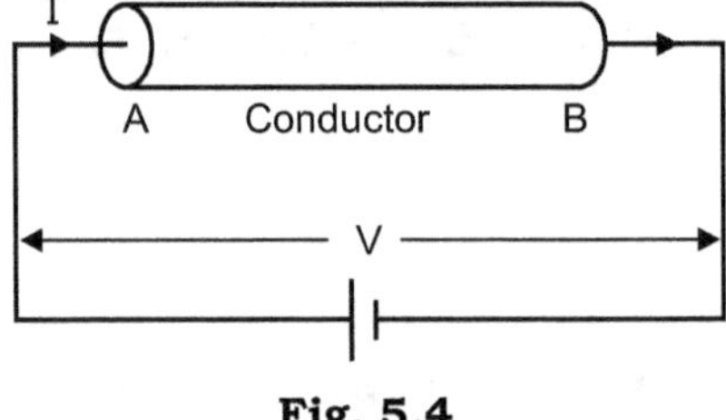

Fig. 5.4

$$V \propto I$$

$$V = \text{Constant} \times I$$

$$\frac{V}{I} = \text{Constant}$$

$$\boxed{\frac{V}{I} = R} \quad \text{or} \quad \boxed{V = IR}$$

5.8 COMBINATION OF RESISTANCES

5.8.1 Resistances in Series

Consider resistances R_1, R_2, and R_3. If they are connected one after the another as shown in Fig. 5.5, then they are said to be connected in series.

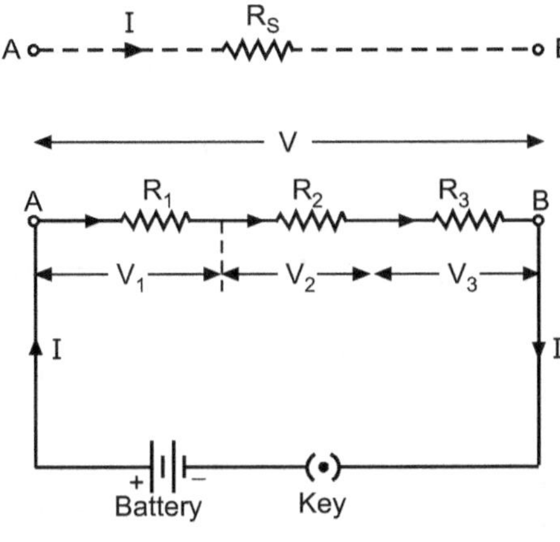

Fig. 5.5

Let series combination of R_1, R_2 and R_3 is connected between points A and B. Potential difference 'V' is applied across the combination.

In series combination, from A to B, there is only one path for flow of current. Therefore, whatever current enters from A, the same current comes out from B i.e for series combination, the current (I) through each of these resistances is same.

But the potential difference 'V' across the combination splits into three parts depending upon the values of R_1, R_2 and R_3.

Let V_1, V_2 and V_3 be the potentials across R_1, R_2 and R_3 respectively.

Thus $\qquad\qquad\qquad V = V_1 + V_2 + V_3$ $\qquad\qquad\qquad\qquad$ ... (5.1)

From Ohm's law, $\qquad V_1 = IR_1 \; ; \; V_2 = IR_2$

$$V_3 = IR_3$$

and $\qquad\qquad\qquad\qquad V = IR_s$

where R_s – equivalent (effective) resistance of series combination

$\therefore$ Equation (5.1) becomes

$$IR_s = IR_1 + IR_2 + IR_3$$

$$I(R_s) = I(R_1 + R_2 + R_3)$$

$$\boxed{R_s = R_1 + R_2 + R_3}$$

In general $R_s = R_1 + R_2 + \ldots\ldots R_n \rightarrow$ for 'n' number of resistances in series.

Law of resistances in series (Statement) : *The law of resistances in series states that the resultant of a number of resistances in series is equal to the sum of the individual resistances. i.e. in short, when number of resistances are connected in series, then the effective resistance increases and will be sum of each resistance.*

5.8.2 Resistances in Parallel

Consider resistances R_1, R_2 and R_3. If they are connected in between two common points as shown in Fig. 5.6 then they are said to be connected in parallel.

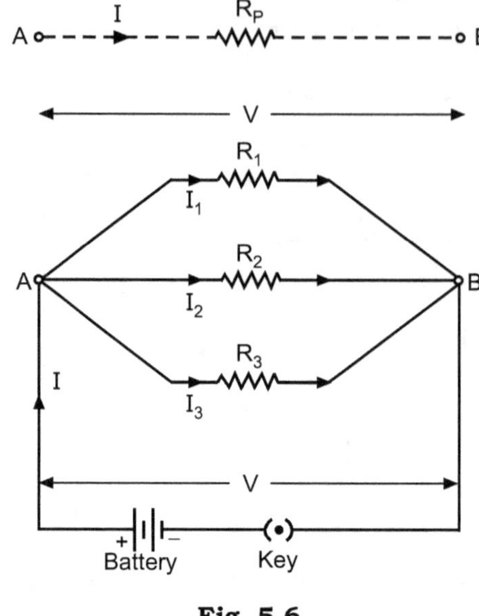

Fig. 5.6

Let parallel combination of R_1, R_2 and R_3 is connected between the points A and B. The potential difference 'V' is applied across the combination.

Since all resistances are connected in between two common points, potential difference across each of them is same.

Current 'I' is flowing through the circuit. At point A this current splits into three parts depending upon the values of R_1, R_2 and R_3 . Let I_1, I_2 and I_3 be the current flowing through R_1, R_2 and R_3 respectively.

Thus $\qquad\qquad I = I_1 + I_2 + I_3 \qquad\qquad\qquad\qquad$ ... (5.2)

From Ohm's law, $\qquad I_1 = \dfrac{V}{R_1} \; ; \; I_2 = \dfrac{V}{R_2} \; ; \; I_3 = \dfrac{V}{R_3}$

$$I = \dfrac{V}{R_p}$$

where, R_p – equivalent (effective) resistance of parallel combination.

$\therefore$ Equation (5.2) becomes,

$$\dfrac{V}{R_p} = \dfrac{V}{R_1} + \dfrac{V}{R_2} + \dfrac{V}{R_3}$$

$$\dfrac{V}{R_p} = V \left(\dfrac{1}{R_1} + \dfrac{1}{R_2} + \dfrac{1}{R_3} \right)$$

$$\boxed{\dfrac{1}{R_p} = \dfrac{1}{R_1} + \dfrac{1}{R_2} + \dfrac{1}{R_3}}$$

In general $\dfrac{1}{R_p} = \dfrac{1}{R_1} + \dfrac{1}{R_2} + ... + \dfrac{1}{R_n}$ $\rightarrow$ for 'n' number of resistances in parallel.

Law of resistances in parallel (Statement) : The law of resistances in parallel states that the reciprocal of resultant of a number of resistances connected in parallel is equal to the sum of reciprocals of individual resistances.

i.e.In short when number of resistances are connected in parallel then the effective resistance decreases and will be less than the least resistance.

5.9 KIRCHHOFF'S LAWS

An electrical circuit is an arrangement of conductors, in which the current can flow around closed paths. The simple circuits can be analysed using Ohm's law and rules for series and parallel combination of resistances. Very often, it is not possible to reduce a circuit to a single loop. Such complicated circuits can be analysed by the use of two simple laws called Kirchhoff's laws stated by Gustav Robert Kirchhoff (1824-1887).

5.9.1 Kirchhoff's First Law (Current Law)

It is stated as, "*the algebraic sum of currents at any junction is equal to zero*".

i.e. $$\sum I = 0$$

While taking algebraic sum of currents, the convention used is : currents approaching the junction are taken as positive and currents leaving the junction are taken as negative.

Consider a junction P in an electrical circuit, as shown in Fig. 5.7. The currents I_1 and I_2 are flowing towards the junction, while currents I_3, I_4 and I_5 are flowing away from the junction. The total current approaching the junction is equal to the total current leaving the junction.

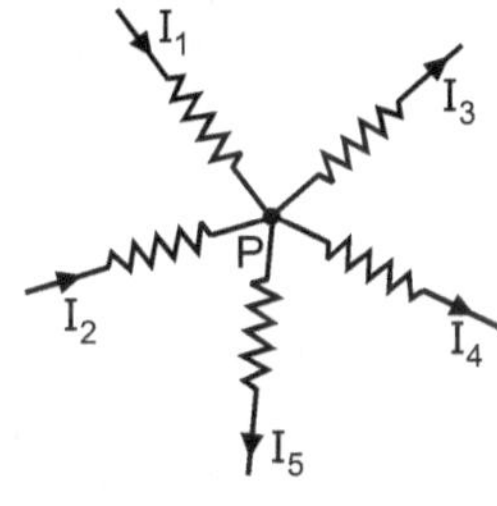

Fig. 5.7

$$\therefore \quad I_1 + I_2 = I_3 + I_4 + I_5$$

$$\therefore \quad I_1 + I_2 - I_3 - I_4 - I_5 = 0$$

$$\therefore \quad \sum I = 0$$

The first law follows from the conservation of charge. Whatever charge enters a given point in unit time in a circuit, must leave that point in the same time because charge is not accumulated at a junction.

5.9.2 Kirchhoff's Second Law (Voltage Law)

It is stated as "*the algebraic sum of e.m.f.s in any closed circuit is equal to the algebraic sum of the products of the current and resistance of each part of that circuit*".

$$\therefore \quad \sum E = \sum IR$$

A closed circuit is a part such that, starting from any point of the circuit, one returns to some point after travelling along all the cells and resistances in that circuit. In a complicated circuit, there may be several closed loops. The law can be applied to each of them. The loop may be traced in clockwise or anticlockwise sense. The sign convention for potential difference used for this law is :

Any potential drop encountered, while going along the loop, is taken as positive and any potential rise is taken as negative.

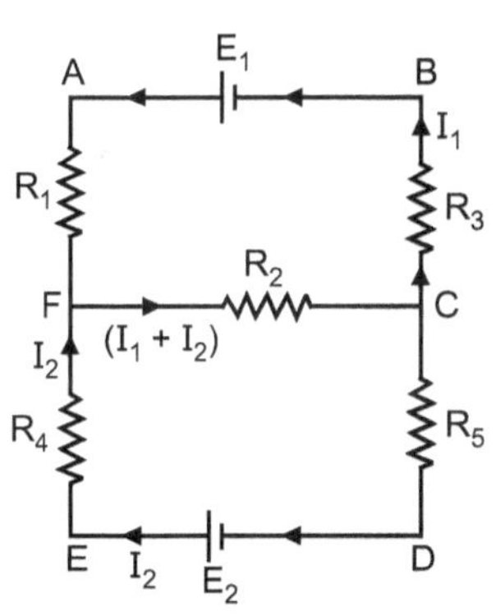

Fig. 5.8

Consider the electrical circuit shown in Fig. 5.8. It consists of three loops ABCFA, ABCDEFA and CDEFC. We can trace the circuit in clockwise sense. Applying the law of the loop ABCDEFA, we get equation,

$$- E_1 - I_1 R_3 + I_2 R_5 + E_2 + I_2 R_4 - I_1 R_1 = 0$$

$$\therefore \qquad E_2 - E_1 = I_1 (R_1 + R_3) - I_2 (R_4 + R_5)$$

$$\therefore \qquad \sum E = \sum IR$$

Similar equations can be written for loops ABCFA and CDEFC and we can determine the unknown currents or potential differences across resistances.

The Kirchhoff's second law is the outcome of the law of conservation of energy applied to an electrical circuit. The product IR is the potential difference across the resistance and it represents the "energy dissipated" in taking a unit charge through the resistance. The e.m.f. E represents the energy supplied by the cell to send a unit charge round the circuit.

$\therefore$ $\sum E$ is the total "energy supplied" to the circuit and $\sum IR$ is the total energy used up in the circuit.

$\therefore$ $\sum E = \sum IR$ explains the law of conservation of energy.

5.10 APPLICATION OF KIRCHHOFF'S LAWS TO WHEATSTONE'S BRIDGE CIRCUIT   (Nov. 18)

One of the most accurate methods of measuring a resistance was developed in 1843 by Charles Wheatstone at Kings college, London. It is known as Wheatstone's bridge (network) method.

Wheatstone's bridge consists of four resistors R_1, R_2, R_3 and R_4 connected to form a closed circuit ABCDA (Fig. 5.9). A source of e.m.f. E is connected across the diagonally opposite points A and C, whereas a galvanometer G is connected across the points B and D.

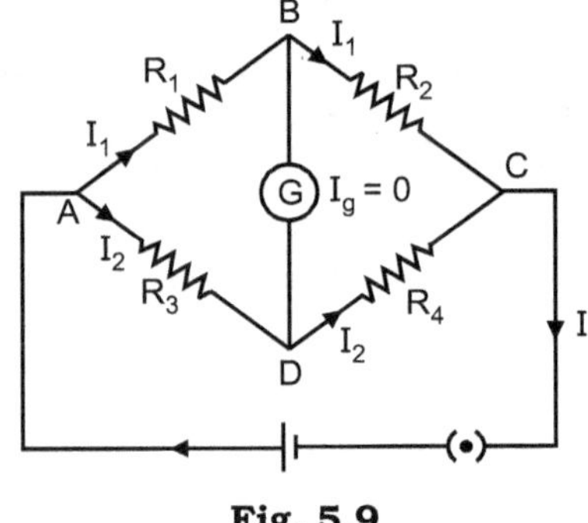

Fig. 5.9

There is fall in potential along the branches ABC and ADC of the network. Potential at B is decided by R_1 and R_2, while potential at D is decided by R_3 and R_4. In general, potentials at B and D viz. V_B and V_D are different. Hence, galvanometer will show deflection. But when $V_B = V_D$, the galvanometer connected between B and D will not show any deflection. Then bridge is said to be balanced. In balanced condition,

$$\frac{R_1}{R_2} = \frac{R_3}{R_4}$$

Proof : Method (1) : The current supplied by a cell is divided in two parts at the junction A, such that I_1 flows through R_1 and I_2 flows through R_3 and in balanced condition, no current flows through the galvanometer. Therefore, current I_1 flows through R_2 and current I_2 flows through R_4.

Let V_A, V_B, V_C and V_D be the potentials at the points A, B, C, and D respectively.

In balanced condition,

$$\text{Potential at B} = \text{Potential at D}$$

$$V_B = V_D$$

$$\therefore \qquad V_A - V_B = V_A - V_D \ \text{ and } \ V_B - V_C = V_D - V_C \qquad \qquad \dots (5.3)$$

Applying Ohm's law to every branch,

$$V_A - V_B = I_1 R_1 \qquad \qquad V_B - V_C = I_1 R_2$$

$$V_A - V_D = I_2 R_3 \qquad \qquad V_D - V_C = I_2 R_4$$

Substituting in equation (5.3), we get

$$I_1 R_1 = I_2 R_3 \qquad \qquad \dots (5.4)$$

$$I_1 R_2 = I_2 R_4 \qquad \qquad \dots (5.5)$$

Dividing equation (5.4) by equation (5.5), we get

$$\frac{R_1}{R_2} = \frac{R_3}{R_4}$$

This is balanced condition of bridge.

Method (2) : Consider above arrangement. Applying Kirchhoff's second law to the loop ABDA, we get

$$I_1 R_1 + I_g G - I_2 R_3 = 0 \qquad \qquad \text{but} \quad I_g = 0$$

$$\therefore \qquad I_1 R_1 = I_2 R_3 \qquad \qquad \dots (5.6)$$

Similarly, by applying the law to the loop BCDB, we get

$$I_1 R_2 - I_2 R_4 + I_g G = 0 \qquad \qquad \text{but} \quad I_g = 0$$

$$\therefore \qquad I_1 R_2 = I_2 R_4 \qquad \qquad \dots (5.7)$$

Dividing equation (5.6) by equation (5.7), we get

$$\frac{R_1}{R_2} = \frac{R_3}{R_4}$$

This is balanced condition of bridge.

If three resistances are known, the remaining resistance can be calculated. Meter bridge and P.O. box also work on the principle of Wheatstone's network.

SUMMARY

- **Electron current :** It is the current due to flow of electrons.

- **Conventional current :** It is the direction of flow of positive charges. It is opposite in direction to that of the electron current.

- **Electric current (I) :** It is defined as the flow of electric charge per unit time.

- **One ampere :** When a charge of 1 coulomb flows through a conductor in 1 sec, then a current of 1 A is said to be flowing.

- **Resistance (R) :** Resistance of material of a wire is defined as the opposition offered by a material to the flow of current. Unit of resistance is ohm shown by a symbol Ω.

- **Resistivity or Specific resistance (ρ) :** It is the resistance of a wire of unit length and unit cross-sectional area. Unit is ohm - meter $\rightarrow \Omega$-m.

- **Conductance (G) :** It is the reciprocal of resistance.

$$G = \frac{1}{R} \dots \text{unit } \frac{1}{\Omega} \text{ or Siemens} \rightarrow S$$

- **Conductivity or Specific conductance (σ) :** It is the reciprocal of specific resistance.

$$\sigma = \frac{1}{\rho} \dots \text{unit } /\Omega m \text{ or S/m}.$$

- **Ohm's law :** It states that, physical conditions of a conductor remaining the same, the electric current flowing through a conductor is directly proportional to the potential difference across it.

$$\boxed{V = IR}$$

- **Kirchhoff's current law :** The algebraic sum of current at any junction is equal to zero. $\therefore \sum I = 0$.

- **Kirchhoff's voltage law :** The algebraic sum of e.m.f.s in any closed circuit is equal to the algebraic sum of products of the current and resistance of each part of that circuit.

$$\Sigma E = \Sigma IR$$

- **Sign convention :** (i) The currents in the direction of tracing the loop are treated as positive, while those in opposite direction are taken as negative (ii) the e.m.f. of cell is positive, if it sends current in the direction of tracing the loop, otherwise negative.

- The balance condition of Wheatstone's network of resistances R_1, R_2, R_3 and R_4 is

$$\frac{R_1}{R_2} = \frac{R_3}{R_4}$$

IMPORTANT FORMULAE

(1) $I = \dfrac{Q}{t}$ where, I = Current in amperes

(2) $V = IR$ Q = Charge in coulomb

(3) $R = \rho \dfrac{L}{A}$ V = Potential difference in volts

(4) Conductance $G = \dfrac{1}{R}$ R = Resistance in ohms

(5) Conductivity, $\sigma = \dfrac{1}{\rho}$ L = Length of a conductor

         A = Cross-sectional area of a conductor

         G = Conductance

(6) $R_s = R_1 + R_2 + R_3$ R_s = Equivalent resistance of series combination

(7) $\dfrac{1}{R_p} = \dfrac{1}{R_1} + \dfrac{1}{R_2} + \dfrac{1}{R_3}$ R_1, R_2, R_3 = Resistances

         R_p = Equivalent resistance of parallel combination

SOLVED EXAMPLES

Example 5.1 : *A 6 m long copper wire has diameter 0.45 mm. If its resistance is 0.6 Ω, calculate its resistivity and conductivity.*

Solution : Given : L = 6 m, dia = 0.45 mm = 0.45×10^{-3} m, $\therefore$ r = 0.225×10^{-3} m,

$$R = 0.6\ \Omega, \quad \text{Resistivity } \rho = ?, \text{ Conductivity } \sigma = ?$$

(i)   Resistivity, $\rho = \dfrac{R(A)}{L} = \dfrac{R\,(\pi r^2)}{L} = \dfrac{0.6 \times [3.142 \times (0.225 \times 10^{-3})^2]}{6} =$

$$\boxed{\rho = 2 \times 10^{-8}\ \Omega m}$$

(ii)   Conductivity, $\sigma = \dfrac{1}{\rho} = \dfrac{1}{(2 \times 10^{-8})}$

$$\boxed{\sigma = 5 \times 10^7\ S/m}$$

Example 5.2 : *A battery of emf 12 volt is connected across a resistance of 10 Ω. Calculate the current flowing through the resistance.*

Solution : Given :   V = 12 V,   R = 10 Ω,   I = ?

We have,   V = IR

$\therefore$ $I = \dfrac{V}{R} = \dfrac{12}{10}$

$$\boxed{I = 1.2\ A}$$

Example 5.3 : *A current of 0.6 A flows through a resistance of 25 Ω. Calculate voltage across it.*

Solution : Given : $I = 0.6$ A, $R = 25$ Ω, $V = ?$

$$V = IR$$

$$V = 0.6 \times 25$$

$$\boxed{V = 15 \text{ V}}$$

Example 5.4 : *A current of 0.8 A flows through a resistance. If battery of emf 12 V is connected across it, calculate the resistance.*

Solution : Given : $I = 0.8$ A, $R = ?$, $V = 12$ V

$$R = \frac{V}{I}$$

$\therefore$ $R = \dfrac{12}{0.8}$

$$\boxed{R = 15 \ \Omega}$$

Example 5.5 : *An electric heater draws a current of 4 A when connected across 220 V supply. What current will it draw when connected across 180 V supply ?*

Solution :

Given :

$I_1 = 4$ A	$I_2 = ?$
$V_1 = 220$ V	$V_2 = 180$ V
$R = \dfrac{V_1}{I_1}$	Now $R = \dfrac{V_2}{I_2}$
$= \dfrac{220}{4}$	$\therefore \quad I_2 = \dfrac{V_2}{R} = \dfrac{180}{55}$
$\boxed{R = 55 \ \Omega}$	$\boxed{I_2 = 3.27 \text{ A}}$

Example 5.6 : *The specific resistance of the material of a cable is 2.81×10^{-7} Ωm. If the resistance of the cable is 2.1 Ω and its radius is 0.8 mm, calculate the length of the cable.*

Solution : Given : $\rho = 2.81 \times 10^{-7}$ Ωm, $R = 2.1$ Ω,

$$r = 0.8 \text{ mm} = 0.8 \times 10^{-3} \text{ m}, \ L = ?$$

We have, $R = \dfrac{\rho L}{A}$

$\therefore$ $L = \dfrac{R\ (A)}{\rho} = \dfrac{R\ (\pi r^2)}{\rho} = \dfrac{2.1 \times [3.142 \times (0.8 \times 10^{-3})^2]}{(2.81 \times 10^{-7})}$

$$\boxed{L = 15.03 \text{ m}}$$

Example 5.7 : *Calculate the resistance of 60 m length of the wire having cross-sectional area of 0.02×10^{-6} m² and having resistivity 3.5×10^{-7} Ωm.*

Solution : Given : $L = 60$ m, $A = 0.02 \times 10^{-6}$ m², $\rho = 3.5 \times 10^{-7}$ Ωm, $R = ?$

We have, $R = \dfrac{\rho L}{A} = \dfrac{(3.5 \times 10^{-7}) \times 60}{(0.02 \times 10^{-6})}$

$$\boxed{R = 1050 \ \Omega}$$

Example 5.8 : *Find the effective resistance of the following combination.*

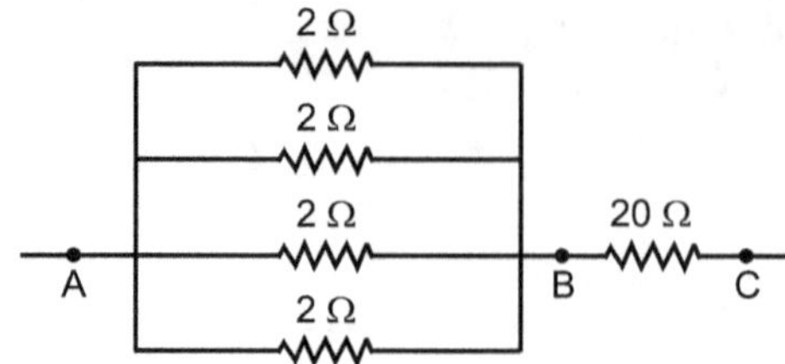

Fig. 5.10

Solution : Let equivalent resistance between AB is R_p.

Thus,
$$\frac{1}{R_p} = \frac{1}{2} + \frac{1}{2} + \frac{1}{2} + \frac{1}{2} = \frac{4}{2} = 2$$

$\therefore$
$$R_p = \frac{1}{2} = 0.5$$

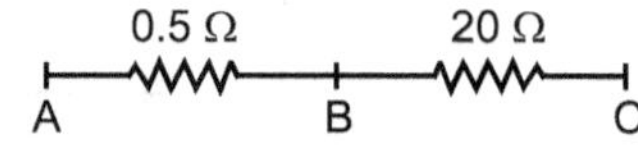

Fig. 5.11

Now, let equivalent resistance between AC is R_s.

$$R_s = 0.5 + 20$$

$\therefore$ $\boxed{R_s = 20.5 \ \Omega}$

Example 5.9 : *A wire of length 240 cm, radius 0.036 cm has a resistance of 12 Ω. Calculate the specific resistance of a wire.*

Solution : Given : $L = 240$ cm $= 2.4$ m, $r = 0.036$ cm $= 0.036 \times 10^{-2}$ m,

$R = 12 \ \Omega$, $\rho = ?$

$$\rho = \frac{RA}{L} = \frac{(R)\,(\pi r^2)}{L} = \frac{(12) \times [3.142 \times (0.036 \times 10^{-2})^2]}{2.4}$$

$$\boxed{\rho = 2.036 \times 10^{-6} \ \Omega\text{-m}}$$

Example 5.10 : *Two resistances have effective resistance of 16 Ω in series and 4 Ω in parallel. Find each resistance.*

Solution : Let R_1 and R_2 be the two resistances.

Given : $R_s = 16 \ \Omega$

$$R_s = R_1 + R_2$$

$$16 = R_1 + R_2 \qquad ...(1)$$

$R_p = 4 \ \Omega$

$$\frac{1}{R_p} = \frac{1}{R_1} + \frac{1}{R_2}$$

$$\frac{1}{R_p} = \frac{R_2 + R_1}{R_1 R_2}$$

$$\therefore R_p = \frac{R_1 R_2}{R_2 + R_1}$$

$$4 = \frac{R_1 R_2}{R_2 + R_1} \qquad ...(2)$$

From (1), $R_2 = (16 - R_1)$. Substitute this value of R_2 in equation (2).

$$\therefore \qquad 4 = \frac{R_1 (16 - R_1)}{(16 - R_1) + R_1}$$

$$\therefore \qquad 4 = \frac{16R_1 - R_1^2}{16}$$

$$\therefore \qquad 64 - 16R_1 + R_1^2 = 0$$

$$(R_1 - 8)(R_1 - 8) = 0$$

$$\therefore \qquad \boxed{R_1 = 8\ \Omega}$$

Substituting this value of R_1 in (1) we get,

$$16 = 8 + R_2$$

$$\therefore \qquad \boxed{R_2 = 8\ \Omega}$$

Example 5.11 : *Calculate the specific resistance of a wire 6 m in length, 0.4 mm in diameter and having a resistance of 30 Ω.*

Solution : Given : $L = 6$ m ; $R = 30\ \Omega$; diameter $= 0.4$ mm

$\therefore$ Radius $r = 0.2$ mm $= 0.2 \times 10^{-3}$ m, Specific resistance $\rho = ?$

$$\rho = \frac{RA}{L} = \frac{R\,(\pi r^2)}{L} = \frac{30 \times [3.142 \times (0.2 \times 10^{-3})^2]}{6}$$

$$\therefore \qquad \boxed{\rho = 63 \times 10^{-8}\ \Omega \text{ - m}}$$

Example 5.12 : *A battery of e.m.f. 12 volts and internal resistance 1 Ω is connected to a resistance of 5 Ω. Calculate the current in the circuit.*

Solution : Given : $E = 12$ V ; $r = 1\ \Omega$; $R = 5\ \Omega$; $I = ?$

$$E = I\,(R + r)$$

$$\therefore \qquad I = \frac{E}{(R + r)} = \frac{12}{(5 + 1)}$$

$$\therefore \qquad \boxed{I = 2\ A}$$

Example 5.13 : *Two resistances of 1 ohm each are connected in parallel. Then 1 ohm is connected in series with the group. Find the resultant resistance. If the current of 10 ampere passes through the circuit, what is the P.D. across each resistance ?*   **(S-95)**

Solution : Given : According to the given information, the circuit diagram is as follows.

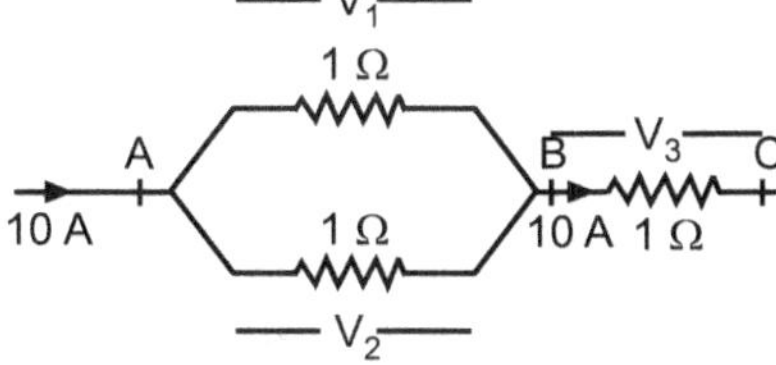

Fig. 5.12

Let resistance across AB is R_p.

$$\therefore \qquad \frac{1}{R_p} = \frac{1}{1} + \frac{1}{1} = \frac{2}{1}$$

$$\therefore \qquad R_p = \frac{1}{2} = 0.5\ \Omega$$

Let resistance across AC is R_s

$$R_s = 0.5 + 1$$
$$R_s = 1.5 \ \Omega$$

Thus resultant resistance = 1.5 Ω

$$V_1 = V_2 = I \times (\text{Resistance between AB}) \qquad \text{... By Ohm's law}$$
$$= I \times (R_p) = 10 \ (0.5)$$

$$V_1 = V_2 = 5 \text{ V}$$

$$V_3 = I \times (\text{Resistance between AC})$$
$$V_3 = 10 \times (1)$$

$$V_3 = 10 \text{ V}$$

Total P.D. acros the combination $V = V_1 + V_2 = 15$ volts

Example 5.14 : *How would you combine four resistances of one ohm each to produce an effective resistance of 0.25 ohm ?*

Solution : Let R_1, R_2, R_3 , R_4 be the given resistances.

Given : $R_1 = R_2 = R_3 = R_4 = 1 \ \Omega$

If they are connected in parallel then,

$$\frac{1}{R_p} = \frac{1}{1} + \frac{1}{1} + \frac{1}{1} + \frac{1}{1} = \frac{4}{1}$$

$$\therefore \qquad R_p = \frac{1}{4} = 0.25 \ \Omega$$

Thus, the effective resistance R_p is 0.25 Ω if the given resistances are connected in parallel.

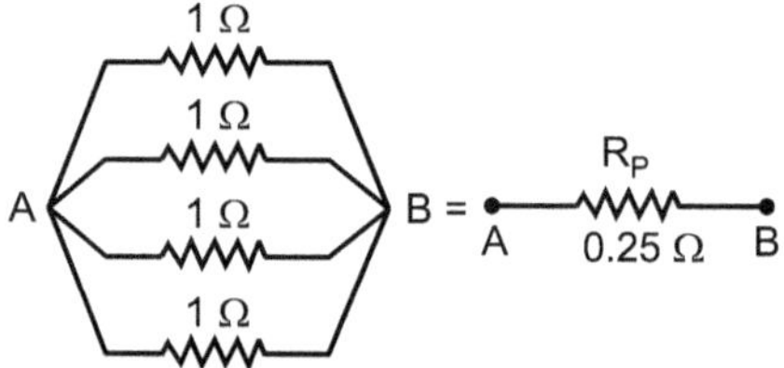

Fig. 5.13

Example 5.15 : *The Wheatstone bridge circuit have the resistance in various arms as shown in Fig. 5.14. Calculate the current through the galvanometer.*

Solution : Let us mark the currents in the various branches according to Kirchhoff's current law. Since there are three unknown quantities I_1, I_2 and I_g, we must consider the following three loops.

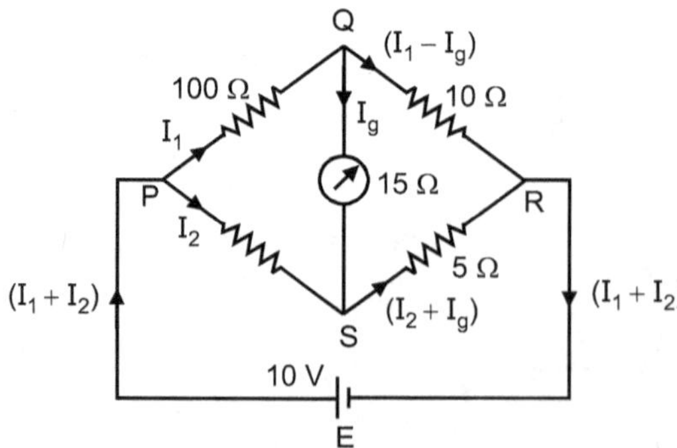

Fig. 5.14

Loop PQSP : Applying Kirchhoff's voltage law to the loop PQSP, we get

$$- 100I_1 - 15I_g + 60I_2 = 0$$

or $\qquad\qquad 20I_1 + 3I_g - 12I_2 = 0 \qquad\qquad\qquad$ ... (1)

Loop QRSQ : Applying Kirchhoff's voltage law to the loop QRSQ, we get

$$- 10\,(I_1 - I_g) + 5\,(I_2 + I_g) + 15I_g = 0$$

or $\qquad\qquad 2I_1 - 6I_g - I_2 = 0 \qquad\qquad\qquad$ ... (2)

Loop PSREP : Applying Kirchhof's voltage law to the loop PSREP, we get

$$- 60I_2 - 5\,(I_2 + I_g) + 10 = 0$$

or $\qquad\qquad 13I_2 + I_g = 2 \qquad\qquad\qquad$ ... (3)

On solving equations (1), (2) and (3), we get,

$$I_g = \textbf{4.87 mA}$$

Example 5.16 : *Determine current in each branch of the network shown in Fig. 5.15.*

Solution : Mark the currents in the various branches according to Kirchhoff's current law. Since are three unknown quantities I_1, I_2 and I_3, we must consider the following three loops.

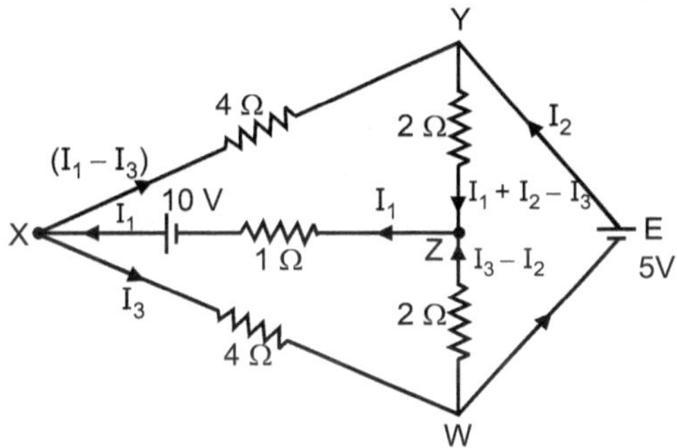

Fig. 5.15

Loop ABCA : Applying Kirchhoff's voltage law to the loop XYZX, we get,

$$- 4\,(I_1 - I_3) - 2\,(I_1 + I_2 - I_3) - 1I_1 + 10 = 0$$

or $\qquad\qquad 7I_1 + 2I_2 - 6I_3 = 10 \qquad\qquad\qquad$ ... (1)

Loop XWZX : Applying Kirchhoff's voltage law to the loop XWZX, we get

$$- 4I_3 - 2\,(I_3 - I_2) - 1I_1 + 10 = 0$$

or $\qquad\qquad I_1 - 2I_2 + 6I_3 = 10 \qquad\qquad\qquad$ ... (2)

Loop XYEWX : Applying Kirchhoff's voltage law to the loop XYEWX, we get,

$$- 4\,(I_1 - I_3) - 5 + 4I_3 = 0$$

or $\qquad\qquad 4I_1 - 8I_3 = - 5 \qquad\qquad\qquad$ ... (3)

Solving equations (1), (2) and (3), we get,

$$I_1 = 2.5 \text{ A}, \ I_2 = 1.875 \text{ A}, \ I_3 = 1.875 \text{ A}$$

Example 5.17 : *Three cells are connected in parallel with their like poles connected together with wires of negligible resistance. If the e.m.f.s of the cells are 2, 1 and 4 V respectively and their internal resistances are 4, 3 and 2 Ω respectively, find the current through each cell.*

Solution : The conditions given in the problem are shown in Fig. 5.16. Let I_1, I_2 and I_3 be the currents through the cells E_1, E_2 and E_3 respectively. Applying Kirchhoff's current law to the junction P, we get,

$$I_1 + I_2 + I_3 = 0$$

or $\qquad\qquad I_3 = - (I_1 + I_2) \qquad\qquad\qquad$ ... (1)

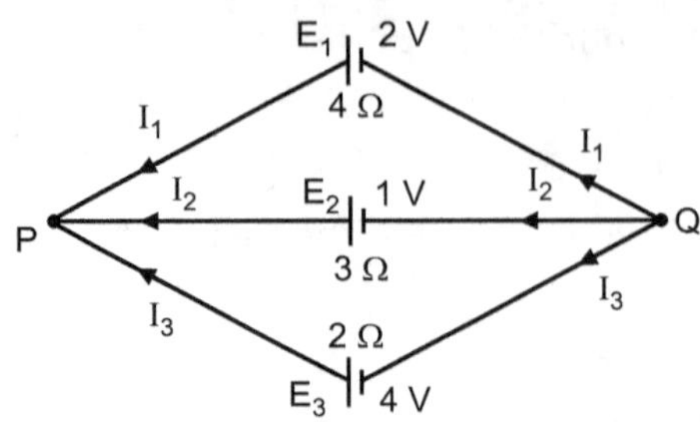

Fig. 5.16

Loop QE₁PE₂Q : Applying Kirchhoff's voltage law to the loop QE₁PE₂Q, we get

$$-4I_1 + 2 + 3I_2 - 1 = 0$$

or $\qquad\qquad 4I_1 - 3I_2 = 1$ ... (2)

Loop QE₁PE₃Q : Applying Kirchhoff's voltage law to the loop QE₁PE₃Q, we get,

$$-4I_1 + 2 + 2I_3 - 4 = 0$$

or $\qquad\qquad 3I_1 + I_2 = -1$ ... (3)

Solving equations (1), (2) and (3), we get

$$I_1 = -\frac{2}{13}\ A;\quad I_2 = -\frac{7}{13}\ A;\quad I_3 = \frac{9}{13}\ A$$

Example 5.18 : *Two batteries E_1 and E_2 having e.m.f.s of 6 V and 2 V respectively and internal resistances of 2 Ω and 3 Ω respectively are connected in parallel across a 5 Ω resistor. Calculate : (i) current through each battery and (ii) terminal voltage.*

Solution : The conditions given in the problem are shown in Fig. 5.17. Let us mark the currents in the various branches according to Kirchhoff's current law. Since there are two unknown quantities I_1 and I_2, we must consider the following two loops.

(i) Loop HBCDEFH : Applying Kirchhoff's voltage law to loop HBCDEFH, we get

$$2I_1 - 6 + 2 - 3I_2 = 0$$

or $\qquad\qquad 2I_1 - 3I_2 = 4$ ... (1)

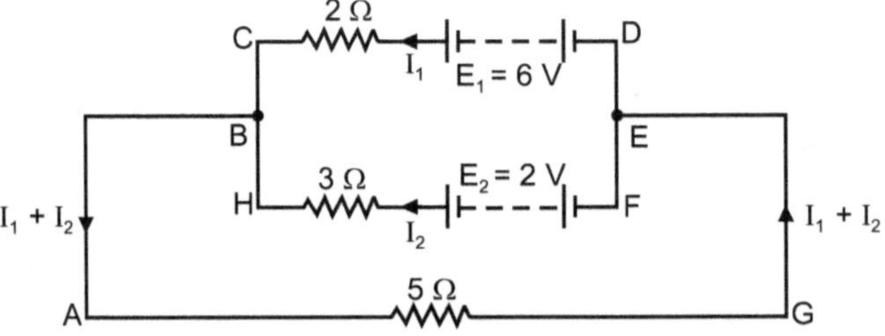

Fig. 5.17

Loop ABHFEGA : Applying Kirchhoff's voltage law to the loop ABHFEGA, we get

$$3I_2 - 2 + 5(I_1 + I_2) = 0$$

or $\qquad\qquad 5I_1 + 8I_2 = 2$ ... (2)

Multiplying equation (1) by 8 and equation (2) by 3 and then adding them, we get

$$31I_1 = 38 \quad\text{or}\quad I_1 = \frac{38}{31} = 1.23\ A$$

i.e., battery E_1 is being discharged at 1.23 A. Substituting $I_1 = 1.23$ A in equation (i), we get,

$$I_2 = -0.52\ A$$

i.e., battery E_2 is charged at 0.52 A.

(ii) $\qquad\qquad$ Terminal voltage $= (I_1 + I_2)\ 5 = (1.23 - 0.52) \times 5 = 3.55\ V$

EXERCISE

1. Define electric current and state its S.I. unit.

2. Define resistance and state its S.I. unit.

3. Define resistivity or specific resistance of a material and state its S.I. unit.

4. Define conductance and state its S.I. unit.

5. Define conductivity or specific conductance and state its S.I. unit.

6. State Ohm's law and state its equation with usual symbol meaning.

7. Derive an expression for equivalent resistance when number of resistances are connected in series.

8. Derive an expression for equivalent resistance when number of resistances are connected in parallel.

9. State law of resistances in series.

10. State law of resistances in parallel.

11. State Kirchhoff's laws of electrical network. State the sign conventions.

12. With the help of a neat circuit diagram, explain Wheatstone's network and obtain the balancing condition.

PROBLEMS FOR PRACTICE

1. A 5 m long wire has diameter 0.4 mm. If its resistance is 10 Ω, calculate its resistivity and conductivity.   **(Ans.** (i) $\rho = 2.5 \times 10^{-7}$ Ωm, (ii) $\sigma = 0.4 \times 10^7$ S/m)

2. A battery of emf 6 V is connected across a resistance of 12 Ω, calculate the current flowing through the resistance.   **(Ans.** I = 0.5 A)

3. A current of 0.8 A flows through a resistance of 30 Ω. Calculate the voltage across it.

    (Ans. V = 24 V)

4. A current of 1.2 A flows through a resistance if a battery of emf 8 V is connected across it. Calculate the resistance.   **(Ans.** R = 6.67 Ω)

5. An electric equipment draws a current of 1 A when connected across 150 V supply. What current will it draw when connected across 220 V supply ?   **(Ans.** I_2 = 1.47 A)

6. The specific resistance of the material of wire is 3×10^{-7} Ωm. If the resistance of the wire is 4 Ω and radius of the wire is 0.6 mm, calculate the length of the wire.   **(Ans.** L = 15.08 m)

7. A wire of length 100 cm, radius 0.15 mm has resistance of 6 Ω. Calculate the specific resistance of the wire.   **(Ans.** 4.23×10^{-7} Ω-m)

8. Find the equivalent resistance of the following system and total current passing through it.

    (Ans. 4 Ω, 3 Amp.)

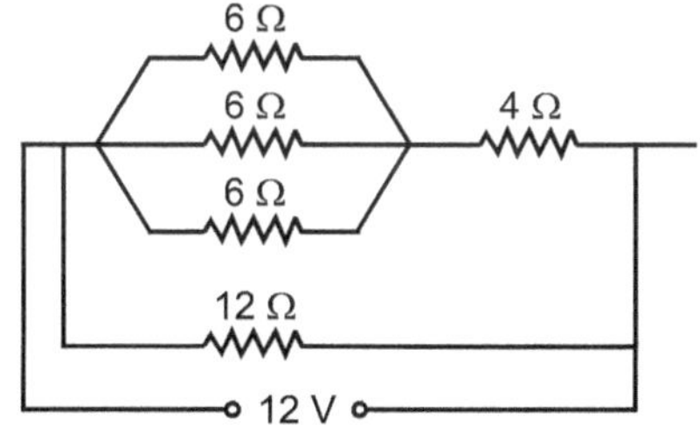

Fig. 5.18

9. When two resistances are connected in series, their effective resistance is 100 Ω, but when they are connected in parallel, the effective resistance becomes 24 Ω. Calculate the two resistances.

(**Ans.** 40 Ω and 60 Ω)

10. With the help of Kirchhoff's laws, find the current in 4 Ω resistor in Fig. 5.19.   (**Ans.** 1.99 A)

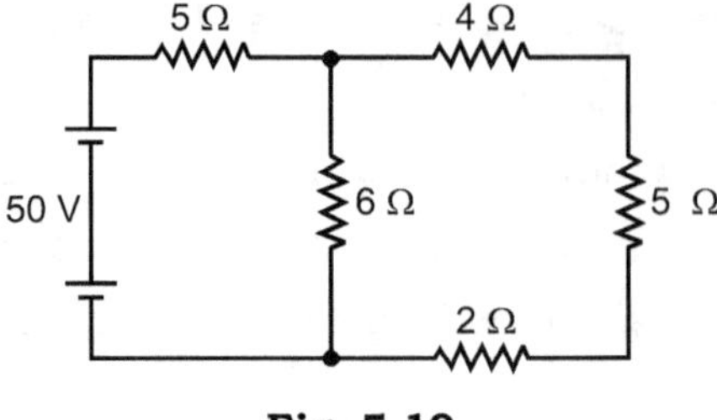

Fig. 5.19

11. In a Wheatstone's bridge network, resistances A = 3 Ω, B = 2 Ω, C = 6 Ω, D = 4 Ω and the galvanometer resistance G = 5 Ω. Calculate the currents in different branches of the network. E.M.F. of the cell is 2 V. Ignore internal resistance of the cell.

(**Ans.** I = 0.6 A, I_A = I_B = 0.4 A, I_C = I_D = 0.2 A, I_G = 0)

12. A current of 0.1 A enters a Wheatstone's bridge consisting of three arms of 10 Ω each and one of 11 Ω. What is the current in the galvanometer if its resistance is 100 Ω ?   (**Ans.** I = 0.221 mA)

13. In a Wheatstone's bridge network A and B, the ratio arms are approximately equal. When C = 500 Ω, the bridge is balanced. On interchanging A and B, the value of C for balance is 505 Ω. Find the value of X and ratio A : B.   (**Ans.** 502.5 Ω, 1 : 1.005)

6

CHAPTER

FIBER OPTICS

6.1   Optical Fiber

6.2   Principle of Optical Fiber or Total Internal Reflection (T.I.R.)

6.3   Propagation in Optical Fiber

6.4   Structure of Optical Fiber (Fiber Optic Construction)

6.5   Fiber Optic Materials

6.6   Critical Angle (θ_c)

6.7   Acceptance Angle (θ_a)

6.8   Numerical Aperture (N_A)

6.9   Types of Optical Fiber

6.10 Transmission Characteristics of Optical Fibers

6.11 Advantages of Optical Fibers in Communication Over Ordinary Cable Communication

6.12 Applications of Optical Fibers

     Summary

     Important Formulae

     Solved Examples

     Exercise

     Problems for Practice

6.1 OPTICAL FIBER

- Light travelling through transparent cylinders by multiple total internal reflection is old and known phenomenon.

- Many experiments were carried out in 1930's to 1950's on glass fiber; which concluded that light can be 'piped' from one point to another point by allowing it to enter one end of the rod of transparent plastic.

- The light will undergo **total internal reflection principle** (i.e. T.I.R. explained in detail later) and follow its contour, emerging at its far end. Images may be transferred from one location to another, using bundle of fine glass fibers.

- When light enters one end of glass fiber under certain right conditions, most of the light propagates.

- In recent three decades, a lot of development has taken place. Currently, all over the world, fiber optics is being used to **transmit voice, television and digital data signals** by light waves over flexible hair-thin threads of glass or plastic.

- Fiber optics is now well proven technology. It is going to influence our lives in a most revolutionary way.

- **Light is faster in speed :** In fiber optics, light is used as signal carrier from one point to another, thus **communication is made faster**.

- **Light has high bandwidth (10 GHz)** i.e. extra information bandwidth, hence it has large capacity to carry signals and messages. Fiber optics communication system has got a tremendous capacity to carry information thousands of times greater than that of electronic communication systems. It carries number of signals through one optical fiber.

- Now a days, we find number of separate cables i.e. telephone cables, T.V. cables, etc. Instead of having number of separate cables, all signals can be carried through a single optical fiber i.e. MANY INTO ONE.

- An optical fiber is a cylindrical tube (pipe) which may be made of glass or plastic or combination. It is manufactured using the principle of TIR in such a way that this tube can guide light from one end to other.

6.2 PRINCIPLE OF OPTICAL FIBER OR TOTAL INTERNAL REFLECTION (T.I.R.)

- Before going into details of T.I.R., we must know about reflection and refraction. When light falls on surface like glass or water, reflection and refraction take place in such a way that this tube can guide light from one end to other.

- Refraction is the property of light on account of which light changes its path (bends), when it enters from one medium into another medium.

- If light enters from air (optically rarer medium) into glass (optically denser medium), the ray bends towards the normal and if light enters from glass to air, the light ray bends away from the normal.

- Also we have seen that as angle of incidence 'i' increases, angle of refraction also increases.

- Maximum possible angle of refraction is 90° and corresponding angle of incidence is called critical angle 'θ_c'. If angle of incidence increases above θ_c, then refraction does not take place (i.e. absorption does not take place) and only reflection takes place.

- **T.I.R. :** Consider light rays in optically denser medium (glass) fall on the surface on the other side of which is less optically denser medium (air) as shown in Fig. 6.1.

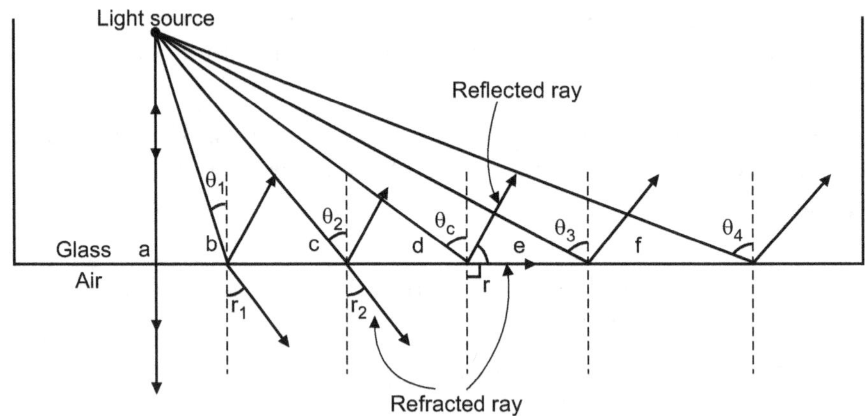

Fig. 6.1

- Consider rays a, b, c in this case, some part of light gets reflected and some part gets refracted.

- As the angle of incidence 'θ' is increased, a situation is reached (see ray 'd') at which the refracted ray points along the surface, and the angle of refraction is 90°.

- For angle of incidence larger than this *critical angle* 'θ_C', no refracted ray exists, giving rise to a phenomenon called *total internal reflection* (T.I.R.).

- Critical angle 'θ_C' is the angle of incidence at which angle of refraction is 90°. Thus for total internal reflection, angle of incidence θ should be greater than θ_C.

- We have, Snell's law of refraction,

$$\frac{\sin i}{\sin r} = \frac{\mu_2}{\mu_1}$$

$$\text{where} \quad \mu_1 = \text{R.I. of first medium}$$

$$\mu_2 = \text{R.I. of second medium}$$

Put $r = 90°$ and $i = \theta_C$

$$\therefore \qquad \frac{\sin \theta_C}{\sin 90°} = \frac{\mu_2}{\mu_1} \qquad \qquad \text{... (6.1)}$$

For example,

For glass and air, $\qquad\qquad \mu_1 = 1.50$ and $\mu_2 = 1.00$

$\therefore$ Equation (6.1) becomes

$$\sin \theta_C = \frac{1.00}{1.50} \qquad \text{since } \sin 90° = 1$$

$$\therefore \qquad \sin \theta_C = 0.667$$

$$\therefore \qquad \theta_C = \sin^{-1}(0.667)$$

$$\therefore \qquad \theta_C = 41.8 \qquad\qquad \text{... Critical angle in case of glass and air}$$

- Total internal reflection does not occur when light originates in the medium of lower index of refraction.

6.3 PROPAGATION IN OPTICAL FIBER

- Now, let us see how light travels in a cable. Fig. 6.2 shows a thin fiber optic cable. A beam of light is focussed on the end of the cable as shown in Fig. 6.2.

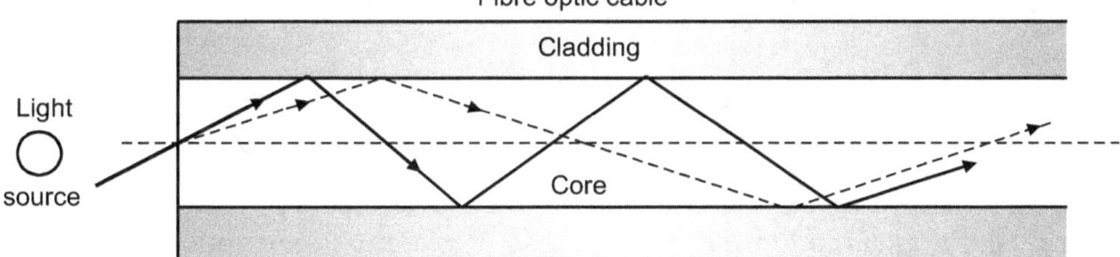

Fig. 6.2

- The angle of incidence of light beam is greater than critical angle. Therefore, **total internal reflection** takes place and the light beams are reflected through the surface of the fiber cables as shown. The beam bounces back and passes the surface cable and it exits at the other end.

- When the light beam reflects from the inner surface of core, the angle of incidence is equal to angle of reflection. Because of this principle, **light rays entering at different angles will take different paths through the cable**.

- Therefore, some light paths will be longer and some will be shorter. Hence, some light rays will exit later and some will exit sooner.

- Because of **total internal reflection**, the light beam will continue to propagate through the fiber eventhough it is bent number of times.

- Only if bending is more which causes the angle of incidence to change which, in turn, will cause the loss of light due to refraction. With long little bends, the light will stay within the cable.

6.4 STRUCTURE OF OPTICAL FIBER (FIBER OPTIC CONSTRUCTION)

- Communication optical fiber has **core** surrounded with **cladding** coated with **protective skin** (insulation, jacket) as shown. Light is transmitted within the core.

- The cladding keeps the light waves within the core because the refractive index of the cladding material is less than that of the core i.e. cladding acts like a reflector and keeps the light within the core based on T.I.R. concept.

- The cladding also provides some strength to the core. The protective skin protects the fiber from moisture. Fiber optic cable is made from either glass or plastic. Another name for glass is silica. Special techniques have been developed and the glass or plastic is melted and pulled such that a fine thread like fiber is formed.

- Glass has superior optical characteristics over plastic.

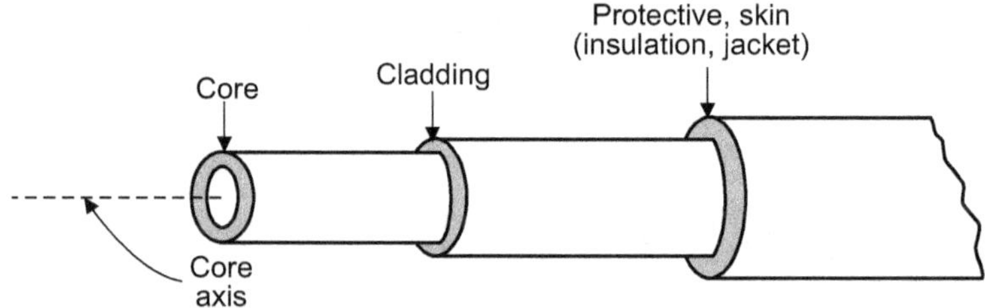

Fig. 6.3 : Structure of an Optical Fiber

- **Dimensions :** The length of the optical fiber is normally 1 km. They can be joined using suitable connectors. Optical fiber is very thin. Its outer diameter ranges from 0.1 mm to 0.15 mm. Core diameter ranges from 5 μm to 600 μm, cladding diameter varies from 125 μm to 750 μm. Thickness of protective skin varies from 30 to 50 μm. To keep light within the core, cladding must have minimum thickness. The protective skin may be thick of 100 μm. Size of core and cladding determines some features of optical fibers.

Conditions :

1. Refractive index of core should be greater than refractive index of cladding.
   i.e. $\mu_{core} > \mu_{cladding}$.

2. Entering light must have angle of incidence (θ) greater than critical angle (θ_c) (certain angle)
   i.e. $\theta > \theta_c$.

6.5 FIBER OPTIC MATERIALS

- The core and cladding is made up of either glass or plastic. Fiber optical cable may be of
  1. Plastic core with plastic cladding.
  2. Glass core with plastic cladding.
  3. Glass core with glass cladding.

- In case of plastic, the core generally of polystyrene and cladding is of silicone or teflon. The glass is made of silica. Also SiO_2 is used.

- Glass (Silica) has superior optical characteristics over plastic. Glass is more expensive than plastic. Plastic is more flexible, but its attenuation of light is greater. Light travels a greater distance in glass than in plastic. For very long distance transmission, glass is preferred. Small amount of boron, phosphorus or germanium is added to adjust refractive index of optical fiber. The refractive index of core and cladding material decide the properties of optical fibers.

- **Material requirements for core :** Core can be made from oxynitride glass or conventional glass. Conventional glass includes silicate glass or fluoride glass, zinc chloride glass, KCl glass. Widely used for infrared, I.R. optical fiber, because of low ionisation and low bulk absorption loss.

- **Material for cladding :** Low loss glass such as silica, bonded polymer, fluoride doped synthetic fused silica.

- **Protective skin (coating) :** Polymide or acrylate coating.

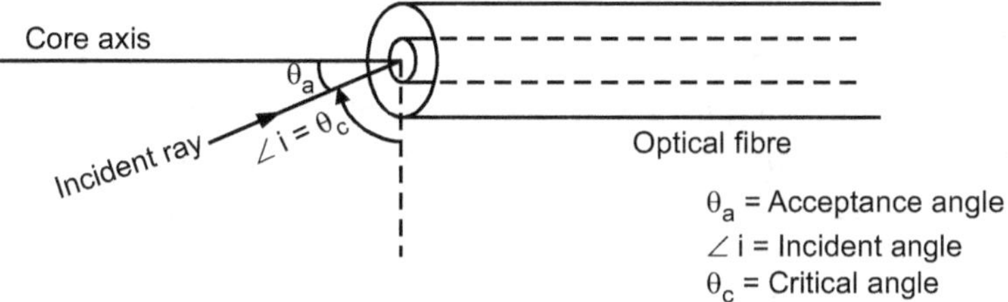

Fig. 6.4

6.6 CRITICAL ANGLE (θ_c) (Nov. 18)

- Critical angle is the angle of incidence at which the angle of refraction is 90°.

OR

- **Critical angle (Definition) :** The critical angle (θ_c) is defined as the minimum angle of incidence above which the total internal reflection will take place.

- Thus, if angle of incidence is greater than critical angle (i.e. $i > \theta_c$), then only reflection takes place (no refraction) and light passes through only core by TIR and no absorption (refraction) in cladding takes place. Thus almost 100% light passes through the core without any loss and signal can be passed from one end of cable to other and communication takes place.

$$\theta_c = \sin^{-1}\left(\frac{\mu_{clad}}{\mu_{core}}\right)$$

6.7 ACCEPTANCE ANGLE (θ_a)

- **The maximum value of external incident angle for which light will propagate in the optical fiber is called the acceptance angle.**

$$\theta_a = \sin^{-1}(N_A) \quad \text{or} \quad \theta_a = \sin^{-1}\sqrt{(\mu_{core}^2 - \mu_{clad}^2)}$$

where, θ_a = acceptance angle

N_A = numerical aperture

- It is also called as acceptance cone half angle.

- It simply means that if an external angle is less than θ_a, TIR takes place and if an external angle is more than it, then light does not propagate upto the far end.

- **Acceptance cone :** We have seen that θ_a is maximum acceptance angle. If θ_a is rotated around the core axis, then the cone is formed which is called as acceptance cone. It is just like a funnel. If light is allowed to fall within this funnel (acceptance cone) then it will propagate upto the far end.

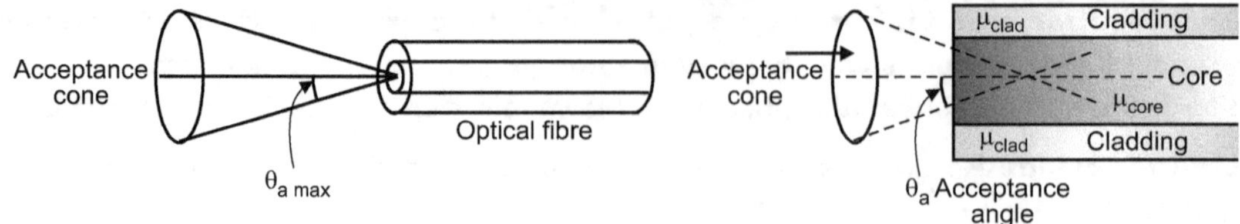

Fig. 6.5

6.8 NUMERICAL APERTURE (N_A)

- It is closely related to acceptance angle. It measures the light gathering power of optical fiber. Greater the magnitude of numerical aperture, greater is the amount of external light the fiber will accept. Numerical aperture is defined as the sine of maximum acceptance angle.

$$N_A = \sqrt{(\mu_{core}^2 - \mu_{clad}^2)}$$

where, N_A = Numerical aperture

μ_{core} = Refractive index of core

μ_{clad} = Refractive index of cladding

Also, $N_A = \sin \theta_a$

where, N_A = Numerical aperture

θ_a = Acceptance angle

6.9 TYPES OF OPTICAL FIBER

Fiber optic cables are classified by two ways :

1. Depending upon the variation in refractive index.

2. Depending upon the mode of propagation. i.e. path that light ray propagate.

Types of optical fiber are :

1. Single mode step index optical fiber.

2. Multimode step index optical fiber.

3. Multimode graded index optical fiber.

Plastic Fiber :

- Used for shorter run.
- Higher attenuation.
- Easy to install.
- Nice withstand stress.
- Less expensive.
- Less weight.

6.9.1 Single Mode Step Index Optical Fiber

- Single mode step index optical fiber carries only one mode. The core is so small and designed for single specific wavelength. Light travelling through this cable follows single path.

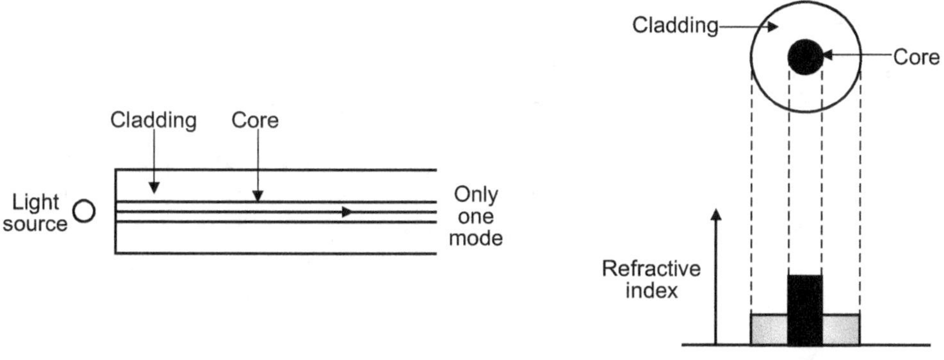

(a) Single mode step index   **(b) Refractive index profile of single mode step index fiber**

Fig. 6.6

- In step index type, material of core has complete uniform distribution of refractive index throughout its thickness. The refractive index of core is greater than that of cladding. Step index refers to the fact that the **refractive index profile suddenly decreases (sharply defined step) from core to cladding**. It simply means that core has one constant refractive index and cladding has another refractive index.

- Because of small core, this type of optical fiber has **low modal dispersion** (low widening tendency). Normally, such a fiber is used for short distance.

- Less attenuation can run over longer distance.

- Larger bandwidth.

6.9.2 Multimode Step Index Optical Fiber

- Multimode step index optical fiber carries many modes. i.e. light travelling through such a fiber takes many paths through the core.

- Step index material of core has complete uniform distribution of refractive index throughout its thickness.

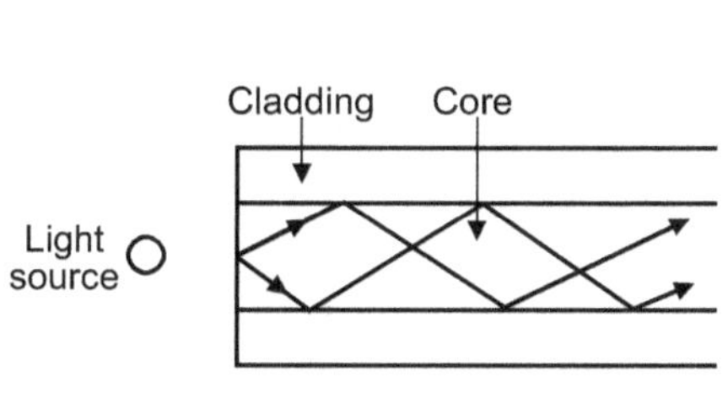

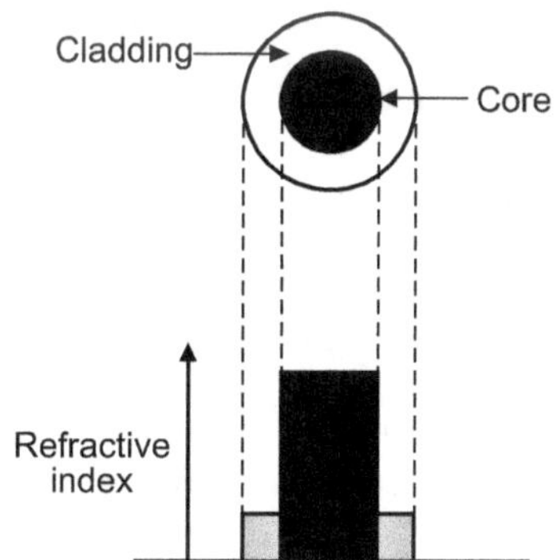

(a) Multimode step index optical fiber   **(b) Refractive index profile of multimode step index fiber**

Fig. 6.7

- The refractive index of core is greater than that of cladding. As we have seen, step index refers to the fact that the **refractive index profile suddenly decreases from core to cladding**.

- In multimode step index fiber, the central core is much larger, **light propagates in zig-zag direction as shown**. There are many paths. As a result, all light rays do not take same time to travel through the length of the fiber.

- Inexpensive, easy to couple light into fiber.

6.9.3 Multimode Graded Index Optical Fiber

- As we have seen (multimode), multimode graded index optical fiber carries many modes. i.e. light travelling through such a fiber takes many paths through the core.

- Graded index central core has non-uniform refractive index. Refractive index of core varies with the radial distance from the axis. The refractive index is maximum at centre (axis) and decreases gradually towards the outer edge as shown in Fig. 6.8 (b).

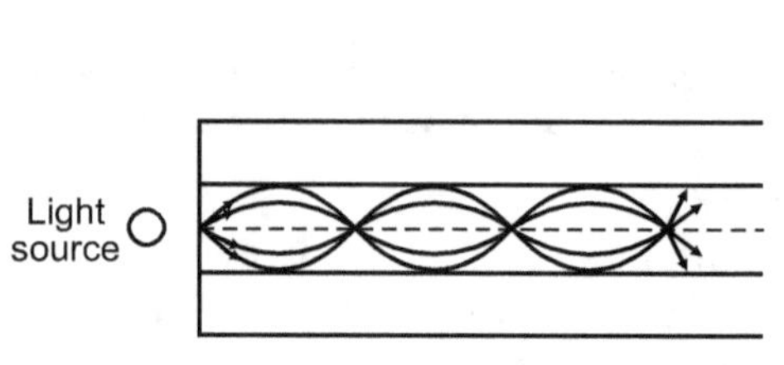

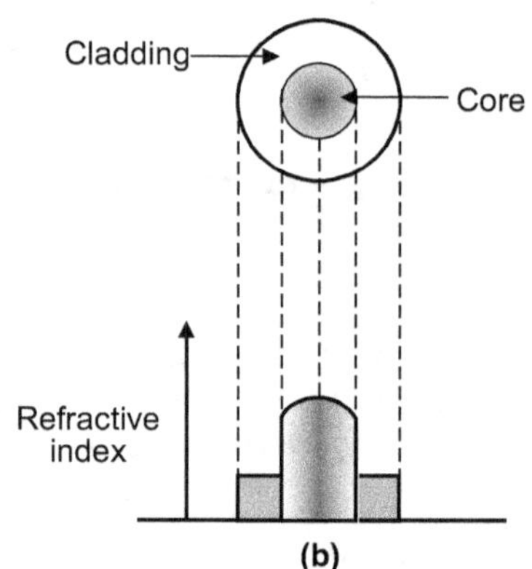

(a) Multimode graded index optical fiber   **(b) Refractive index profile of multimode graded index fiber**

Fig. 6.8

- Because of graded variation in refractive index across the core, the light rays bend smoothly. **The light propagates in curved fashion**.

- The light rays near the edge of the core take a longer path, but travel faster since the index of refraction is lower. Then all the modes (light paths) tend to arrive at one point simultaneously.

- The result is that it has **less modal dispersion**.

6.10 TRANSMISSION CHARACTERISTICS OF OPTICAL FIBERS

- The study of transmission characteristics of optical fibers helps in selection of the fiber.

1. **Absorption loss :** Ultraviolet absorption takes place for pure fused silica. Valence electrons can be ionised. This ionization amounts loss of energy.

   Infrared absorption because of photons of light energy are absorbed by atoms within the glass molecule and convert into vibrational (heat) energy.

2. **Material or Reyleigh scattering loss :** It is due to variation in refractive index in fiber during manufacturing. Because of this diffraction takes place.

3. **Dispersion :** It takes place because of widening of light ray inside the fiber.

4. **Attenuation :** It is the power loss which takes place inside the cable. The main sources which are responsible for attenuation are electron absorption, material absorption, impurity absorption.

5. **Bending loss :** Bending provides scattering of light rays.

6. **Coupling loss :** It is due to misalignment between two cables and due to core size mismatch.

6.11 ADVANTAGES OF OPTICAL FIBERS IN COMMUNICATION OVER ORDINARY CABLE COMMUNICATION

1. In case of optical fiber since **light** is a signal carrier, communication is **speedy**. In case of ordinary cable communication, **electricity** is signal carrier.

2. Optical fiber communication is electrically isolated since light carries signal i.e. safe data transfer takes place speedily.

3. Light has high bandwidth (10 GHz) i.e. extra information bandwidth, hence many signals can be sent through single fiber. But in case of ordinary cable for every signal, there may be different cables required. i.e. in future we may not find mesh of number of cables from terrace to terrace.

4. **Lighter weight :** Optical fibers because of their light weight and flexibility can be handled very easily than that of heavy copper cables.

5. **Opposition to inductive interference :** Ordinary cables are metallic, hence they pick up electromagnetic waves which interfere and may be cause of distortion. But optical fiber since they are not metallic, do not pick up electromagnetic waves.

6. Longer life, easy maintenance, temperature resistant.

7. **Cross talk immunity :** There is no signal leakage and hence cross talk between neighbouring fibers are almost absent.

6.12 APPLICATIONS OF OPTICAL FIBERS

1. To carry telephone signals in telephone department.

2. To carry telex signals.

3. To carry FAX signals (facsimile sender).

4. Used in internet to carry signals.

5. To carry T.V. cable signals.

6. In medical instrumentation.

7. It is used in ship, aircraft.

8. In military it is used in land-based system, airborne system, naval system, undersea system etc.

9. It has applications in space, computer, industry and also in photoreceptor optics.

SUMMARY

- Properties of light i.e. high speed, high bandwidth, electrically isolated play important role in optical fiber.

- Total internal reflection is the principle of optical fiber.

- Optical fiber structure consists of inner core, cladding and protective skin.

- Maximum external incident angle for which light will propagate in the optical fiber is called acceptance angle.

- Numerical aperture measures light gathering power of optical fiber.

- There are three types of optical fiber, which are (1) single mode step index, (2) multimode step index, (3) multimode graded index optical fiber.

Single mode step index fiber	Multimode step index fiber	Multimode graded index fiber
1. This type of fiber carries single mode.	1. This type of fiber carries many modes.	1. This type of fiber carries many modes.
2. Core diameter is very small.	2. Core diameter is comparatively large.	2. Core diameter is comparatively large.
3. Refractive index of core is uniform throughout its thickness.	3. Refractive index of core is uniform throughout its thickness.	3. Refractive index of core varies with radial distance from the axis. The refractive index is maximum at axis of core and decreases gradually towards the older edges.
4. Less modal dispersion.	4. More modal dispersion.	4. Less modal dispersion.
5. Light propagates in linear fashion.	5. Light propagates in zig-zag fashion.	5. Light propagates in curved fashion.

IMPORTANT FORMULAE

1. $\theta_c = \sin^{-1}\left(\dfrac{\mu_{clad}}{\mu_{core}}\right),$ where θ_c = Critical angle

2. $N_A = \sqrt{(\mu_{core}^2 - \mu_{clad}^2)},$ μ_{clad} = Refractive index of cladding

3. $N_A = \sin\theta_a$ μ_{core} = Refractive index of core

4. $\theta_a = \sin^{-1}(N_A)$ N_A = Numerical aperture

    $= \sin^{-1}\left(\sqrt{(\mu_{core}^2 - \mu_{clad}^2)}\right)$ θ_a = Acceptance angle

SOLVED EXAMPLES

Numericals on Critical angle, Numerical aperture, Acceptance angle :

Example 6.1 :

For silicate glass optical fiber, calculate the critical angle if refractive index of core is 1.55 and refractive index of cladding is 1.35.

Solution :

Given : $\mu_{core} = 1.55$, $\mu_{clad} = 1.35$, $\theta_c = ?$

We have, $\theta_c = \sin^{-1}\left(\dfrac{\mu_{clad}}{\mu_{core}}\right)$

               $= \sin^{-1}\left(\dfrac{1.35}{1.55}\right)$

$$\boxed{\theta_c = 60.57°}$$

Example 6.2 :

Calculate numerical aperture of an optical fiber having refractive index of core as 1.5 and refractive index of cladding as 1.3.

Solution :

Given :

$$N_A = ?$$
$$\mu_{core} = 1.5$$
$$\mu_{clad} = 1.3$$

We have,

$$N_A = \sqrt{(\mu_{core}^2 - \mu_{clad}^2)}$$
$$= \sqrt{(1.5^2 - 1.3^2)}$$
$$N_A = \sqrt{0.56}$$
$$\boxed{N_A = 0.748}$$

Example 6.3 :

Calculate acceptance angle required in an optical fiber having numerical aperture as 0.75.

Solution :

Given : $\theta_a = ?,$ $N_A = 0.75$

We have,

$$\theta_a = \sin^{-1}(N_A)$$
$$= \sin^{-1}(0.75)$$
$$\boxed{\theta_a = 48.59°}$$

Example 6.4 :

Calculate acceptance angle required for an optical fiber having refractive index of core as 1.55 and refractive index of cladding as 1.35.

Solution : Given :

$$\theta_a = ?$$
$$\mu_{core} = 1.55$$
$$\mu_{clad} = 1.35$$

We have,

$$\theta_a = \sin^{-1}(N_A)$$
$$= \sin^{-1}\left(\sqrt{(\mu_{core}^2 - \mu_{clad}^2)}\right)$$
$$= \sin^{-1}(\sqrt{1.55^2 - 1.35^2})$$
$$= \sin^{-1}(0.76)$$
$$\boxed{\theta_a = 49.6°}$$

Example 6.5 :

Calculate numerical aperture and acceptance angle for an optical fiber.

Given : R.I. of core = 1.40, R.I. of cladding = 1.35.

Solution : Given :

$$\mu_{core} = 1.40$$
$$\mu_{clad} = 1.35$$
$$N_A = ?$$
$$\theta_a = ?$$

We have,
$$N_A = \sqrt{(\mu_{core}^2 - \mu_{clad}^2)}$$
$$= \sqrt{(1.40)^2 - (1.35)^2} = \sqrt{(1.96 - 1.8225)}$$
$$= \sqrt{0.1375}$$
$$\boxed{N_A = 0.371}$$

Acceptance angle,
$$\theta_a = \sin^{-1}(N_A) \text{ or } \sin^{-1}\left(\sqrt{\mu_{core}^2 - \mu_{clad}^2}\right)$$
$$= \sin^{-1}(0.371)$$
$$\boxed{\theta_a = 21.77°}$$

Example 6.6 :

An optical fiber has a numerical aperture of 0.2, a core of refractive index 1.40.

Calculate (i) Refractive index of cladding, (ii) The acceptance angle of a fiber.

Solution :

  Given :
$$N_A = 0.2$$
$$\mu_{core} = 1.40$$
$$\mu_{clad} = ?$$
$$\theta_a = ?$$

We have,
$$N_A = \sqrt{(\mu_{core}^2 - \mu_{clad}^2)}$$
$$0.2 = \sqrt{(1.40^2 - \mu_{clad}^2)}$$

Squaring on both sides, we get
$$0.04 = 1.96 - \mu_{clad}^2$$
$$\mu_{clad}^2 = 1.96 - 0.04$$
$$\mu_{clad}^2 = 1.92$$
$$\mu_{clad} = \sqrt{1.92}$$
$$\boxed{\mu_{clad} = 1.386}$$

Now,
$$N_A = \sin \theta_a$$
$$\therefore \quad \theta_a = \sin^{-1} N_A$$
$$\theta_a = \sin^{-1}(0.2)$$
$$\therefore \quad \boxed{\theta_a = 11.54°}$$

Example 6.7 :

Numerical aperture of a fiber is 0.244 and refractive index of cladding is 1.48. Calculate refractive index of core and acceptance angle.

Solution :

  Given :
$$N_A = 0.244$$
$$\mu_{clad} = 1.48$$
$$\mu_{core} = ?$$
$$\theta_a = ?$$

We have,
$$N_A = \sqrt{\mu_{core}^2 - \mu_{clad}^2}$$
$$0.244 = \sqrt{\mu_{core}^2 - 1.48^2}$$

Squaring on both sides,

$$(0.244)^2 = \mu_{core}^2 - (1.48)^2$$

$$\therefore \quad \mu_{core}^2 = 0.244^2 + (1.48)^2$$

$$\mu_{core}^2 = 2.249$$

$$\mu_{core} = \sqrt{2.249}$$

$$\boxed{\mu_{core} = 1.499}$$

$$\theta_a = \sin^{-1}(N_A)$$

$$= \sin^{-1}(0.244)$$

$$\boxed{\theta_a = 14.12°}$$

EXERCISE

1. State the basic principle of optical fiber.

2. Define critical angle, acceptance angle and numerical aperture.

3. State three types of optical fiber.

4. Explain single mode step index optical fiber.

5. Explain multimode step index optical fiber.

6. Explain graded mode index optical fiber.

7. What are the advantages of optical fiber over ordinary cable communication ?

8. State applications of optical fiber.

9. Distinguish between : (a) single mode step index fiber, (b) multimode step index fiber, (c) multimode graded index optical fiber.

10. Write two differences between multimode step index fiber and multimode graded index fiber.

11. State any two points to distinguish between (i) step index and (ii) graded index optical fiber.

12. Draw the configuration of optical fiber.

13. State the conditions for TIR.

14. What is numerical aperture ? Explain with labelled diagram. State formula for numerical aperture.

15. What are the types of an optical fiber ? Explain propagation of light through an optical fiber.

16. State the principle of optical fiber. Give its construction and write conditions that must be satisfied for working of optical fiber.

17. Mention types of optical fiber and explain any one.

PROBLEMS FOR PRACTICE

1. For a glass optical fiber, calculate the critical angle if refractive index of core is 1.5 and refractive index of cladding is 1.3. **(Ans.** $\theta_c = 60.07°$)

2. Calculate numerical aperture of an optical fiber having refractive index of core as 1.57 and refractive index of cladding as 1.32. **(Ans.** $N_A = 0.85$)

3. Calculate acceptance angle required for an optical fiber having numerical aperture 0.83.

 (Ans. $\theta_a = 56.099°$)

4. Calculate acceptance angle required for an optical fiber having refractive index of core as 1.48 and refractive index of cladding as 1.36. **(Ans.** $\theta_a = 35.72°$)

5. Calculate numerical aperture and acceptance angle for an optical fiber.

 (Given : R.I. of core = 1.47, R.I. of cladding = 1.33) **(Ans.** $N_A = 0.626$, $\theta_a = 38.76°$)

6. An optical fiber has a numerical aperture of 0.37, a core of refractive index 1.49. Calculate :

 (i) Refractive index of cladding.

 (ii) The acceptance angle of an optical fiber. **(Ans.** $\mu_{clad} = 1.443$, $\theta_a = 21.72°$)

7. Numerical aperture of a fiber is 0.65 and refractive index of cladding is 1.29. Calculate refractive index of core and acceptance angle. **(Ans.** $\mu_{core} = 1.44$, $\theta_a = 40.54°$)

7

CHAPTER

BAND THEORY OF SOLIDS

7.1 Introduction

7.2 Energy Band Diagrams for Conductors, Semiconductors and Insulators

7.3 Intrinsic and Extrinsic Semiconductors

7.4 P-N Junction Diode - Forward and Reverse Biased Characteristics

Summary

Exercise

7.1 INTRODUCTION

- For proper understanding of band theory, we need to study the atomic structure and electron energies in an atom.

Atomic structure :

- An atom is the fundamental unit of matter and it has an independent existence. An atom consists of a centrally placed heavy nucleus.

- Electrons are revolving in elliptical orbit around the nucleus. Nucleus consists of protons and neutrons. Protons carry positive charge, whereas neutrons are electrically neutral. Electrons carry negative charge.

- Electrons are much lighter in weight than protons or neutrons. Since matter in the normal state is electrically neutral, the atom should also be neutral i.e. an atom in its normal state consists of equal number of electrons and protons.

- Electrons in copper and electrons in iron are same. But the difference is that the number of electrons and their arrangement is different.

- Electrons of an atom do not move in the same orbit. They move in different orbits or shells. The maximum number of electrons that can exist in the orbit = $2n^2$, where 'n' is the number of orbits or shells.

  i.e. First shell consists of $2(1)^2 = 2$ electrons.

  Second shell consists of $2(2)^2 = 8$ electrons.

  Third shell consists of $2(3)^2 = 18$ electrons.

- But there is an exception to this rule and that is, the outermost orbit in an atom cannot accommodate more than 8 electrons.

- The electrons in the inner shell do not normally leave the atom. Electrons which are in the outer shell, revolve with high speed and do not always remain restricted to the same atom. Some of them move in a random manner and travel from one atom to another atom. Such electrons are called free electrons.

- The electrons in the inner shell are bound to the nucleus and hence called bound electrons. The electrons in the outermost orbit are called valence electrons.

Electron energies :

- In an atom, various electrons are distributed in various shells. Electrons do not have any arbitrary value of energy in an orbit, but only certain permissible values.

- No electron can exist at an energy level apart from this permissible one. Fig. 7.1 shows different energy levels available for an electron in a single atom.

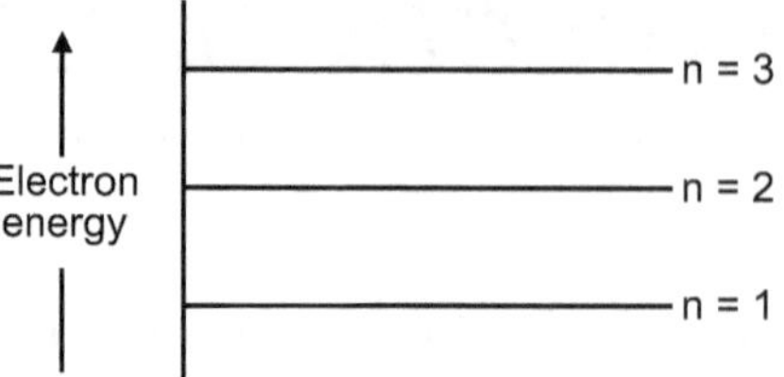

Fig. 7.1

- As we have seen, electron orbiting very close to the nucleus is bound to the nucleus and possesses very small amount of energy. Therefore, it is difficult to remove such an electron. On the other hand, an electron orbiting in the outer shell has more energy and it could easily be removed from its orbit. These electrons are called as valence electrons. These valence electrons take part in the chemical reaction and also in bonding the atoms together.

7.1.1 Formation of Energy Band in Solids

- When large number of atoms are brought together, the atoms experience interatomic interactions. i.e. in a solid, the orbit of an electron is influenced not only by the charges in its own atom, but by nuclei and electrons of neighbouring atoms.

- Thus, the electrons in the outermost shell are forced to have energies different from those in the atoms.

- **Since each electron occupies a different position inside the solid, no two electrons have same pattern of surrounding charges. Thus, orbits of the electrons are different**.

- **There are millions of electrons belonging to the first orbit of atoms in the solid. Each of these electrons has different energy. The different energy levels are closely packed. Since there are millions of first-orbit electrons, the closely spaced energy level forms a cluster or continuous band of energy. Similarly, the second-orbit electrons form second band**.

- Thus, if an atom is considered separately, then electron in any orbit has fixed energy. But in a solid, many such atoms are closely packed and hence each electron in any orbit is under the influence of this closely packed structure.

- Because of this, instead of fixed energy, the electrons have a range of energies i.e. energy bands.

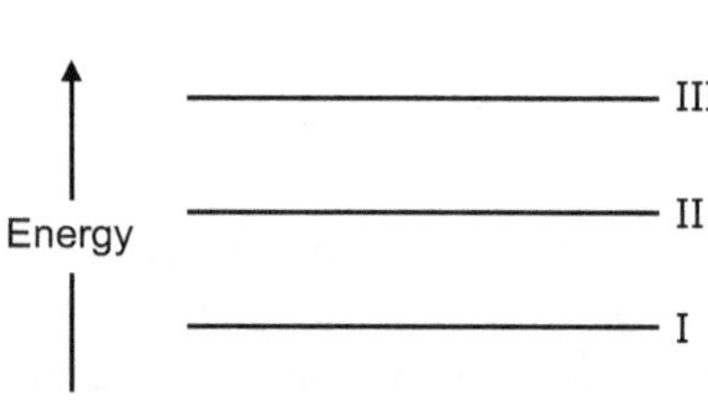

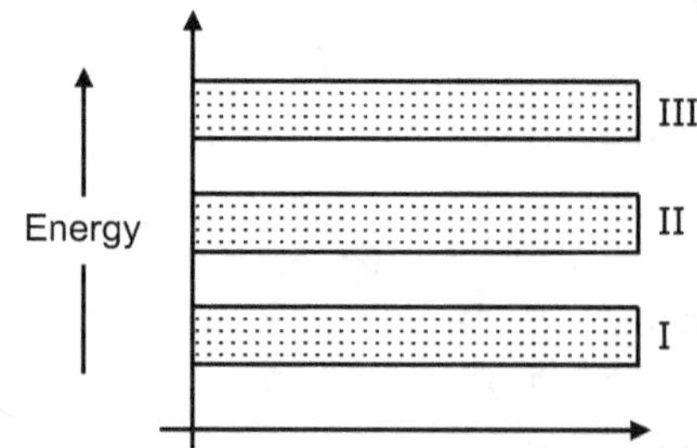

Fig. 7.2 : Fixed energy levels of separate atoms

Fig. 7.3 : Range of energies (band) possessed by an atom when a solid is considered

- There are number of energy bands in a solid. Following are the important energy bands.

Valence band :

- The electrons in the outermost orbit of an atom are known as valence electrons. For a normal atom, the valence electrons have highest energy.

- For inert gases, valence band is completely filled. For other elements, it may be completely filled or partially filled with electrons.

- **The range of energies possessed by valence electrons is known as the valence band.**

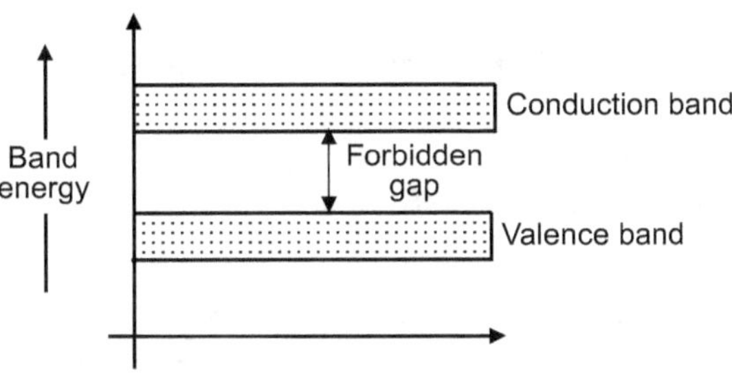

Fig. 7.4

Conduction band :

- For metals, the valence electrons are loosely attached to the nucleus. So they can be detached and become free electrons. These free electrons are responsible for carrying current in a conductor. Hence, they are called **conducting electrons**. **The range of energies possessed by conducting electrons is known as the conduction band.** It may be either empty or partially filled with electrons.

Forbidden energy gap :

- There is an energy gap, E_G, (energy difference) between the valence band and conduction band. An electron can be lifted from valence band to conduction band by adding energy 'E_G'. If we add energy less than E_G, material will not accept it because no permissible energy level exists between valence band and conduction band. Therefore, the gap between valence band and conduction band is called as **forbidden gap**.

- For insulators, this forbidden gap is very large and for semiconductors, it is very small. For conductors, there is no such forbidden gap, and the valence band and conduction band overlap each other.

7.2 ENERGY BAND DIAGRAMS FOR CONDUCTORS, SEMICONDUCTORS AND INSULATORS

- Solids are classified as insulators, conductors and semiconductors, which can be explained with the help of energy band diagram as follows.

7.2.1 Conductor

- A material conducts electricity if it contains movable charges in it. The free electrons move randomly inside a solid, and carry charge. Thus, free electrons are charge carriers.

- Metal such as copper contains large number of free electrons at room temperature. Here, conduction band and valence band overlap each other i.e. valence band energies are same as the conduction band energies.

- Thus, the band structure of conductor (metal) contains no forbidden energy gap. Thus, a very small potential across the conductor makes the free electrons to give rise to an electric current. e.g. silver, copper, aluminium.

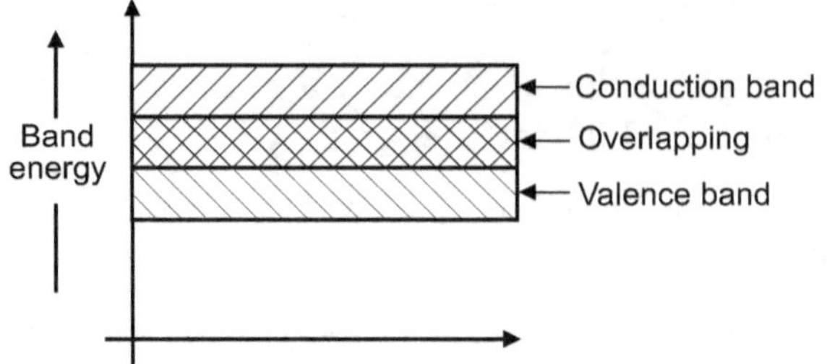

Fig. 7.5 : Energy band diagram for conductors (metals)

7.2.2 Semiconductors

- Conductivity of semiconductor material is less than that of conductor and more than that of insulator e.g. silicon, germanium.

- In this case, the valence band is full and conduction band is almost empty. The forbidden energy gap is very small. It is of the order of 1 eV. (For germanium, E_G = 0.72 eV and for silicon, E_G = 1.12 eV).

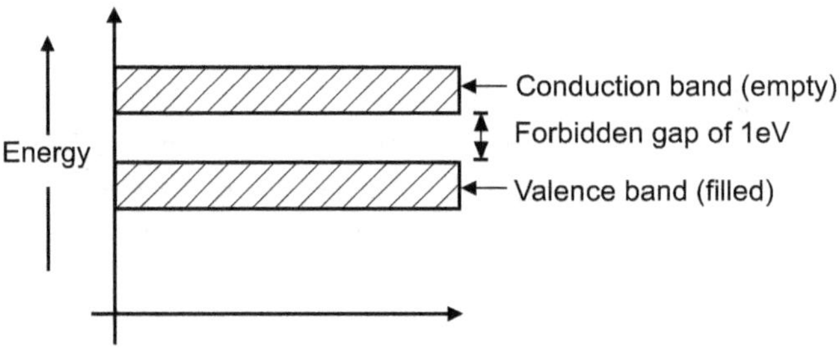

Fig. 7.6

- Thus, even heat energy at room temperature is sufficient to lift the electrons from valence band to conduction band. Some electrons jump from valence band to conduction band. Thus, at room temperature, it conducts some electric current.

- With increase in temperature, width of forbidden energy band is decreased, so that some of the electrons are liberated into the conduction band. In other words, conductivity of semiconductor increases with temperature.

7.2.3 Insulator

- In case of an insulator, the gap between the valence band and conduction band i.e. forbidden energy gap is very wide. E_G is greater than 5.5 eV.

- Because of this, it is practically not possible for an electron in the valence band to jump in the conduction band. So these materials work as insulator. e.g. Glass, wood, rubber, plastic, etc.

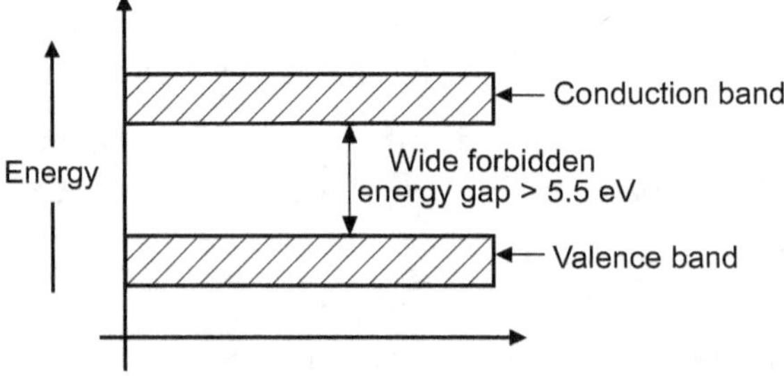

Fig. 7.7

- Semiconductors are classified as :

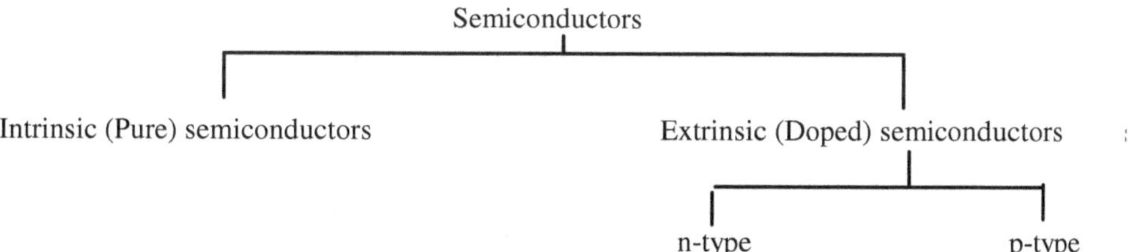

7.3 INTRINSIC AND EXTRINSIC SEMICONDUCTORS

7.3.1 Intrinsic Semiconductor

- A semiconductor such as silicon or germanium in its purest form is called as an **intrinsic semiconductor**.

Atomic bonding :

- An intrinsic semiconductor like pure silicon or germanium has **four** electrons in its outermost orbit of its atoms.

- In order to fill the valence shell, each atom requires four more electrons, which is done by **sharing** one electron from each of the four neighbouring atoms as shown in Fig. 7.8.

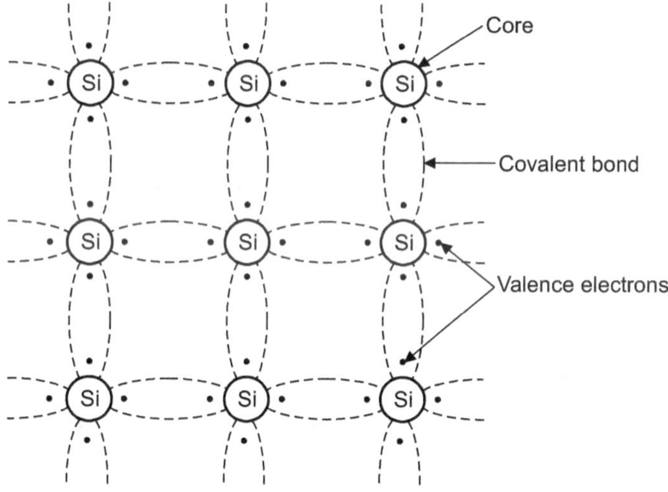

Fig. 7.8 : Crystalline structure of silicon (semiconductor)

- The atoms arrange themselves such that each atom is surrounded by four atoms. This pattern is known as a crystal.

- A covalent bond consists of two electrons one from each adjacent atom. Both electrons are shared by two atoms.

- At absolute zero temperature, all valence electrons are tightly bound to the parent atoms. Thus, there are no free electrons available for electrical conduction. **Thus, semiconductor behaves as a perfect insulator at absolute zero temperature.**

- Heat at room temperature is sufficient to break few of the covalent bonds. Thus, few electrons become free to move in the crystal as shown in Fig. 7.9.

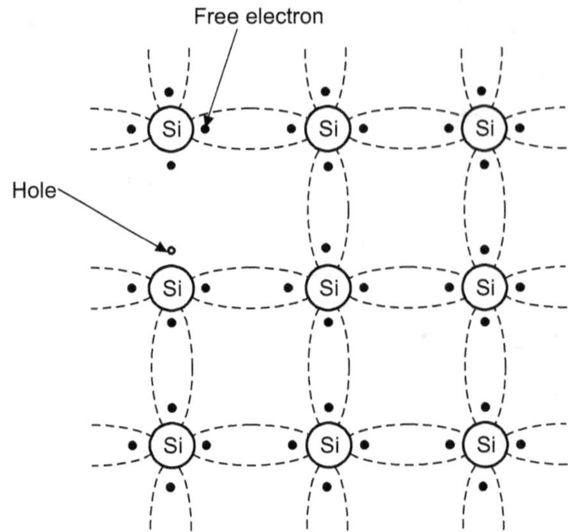

Fig. 7.9 : Generation of electron-hole pair in an intrinsic semiconductor (Si crystal)

- When electron moves away from broken covalent bond, a vacancy i.e. hole is created in the broken band. This vacancy is called hole (i.e. short of electron). Hole is a short of electron, hence it is taken as a positive charge.

- Whenever free electron is generated, a hole is created simultaneously. This generation of electron-hole pair is called thermal generation. The amount of energy required to break a covalent bond is 0.72 eV in case of germanium and 1.12 eV in case of silicon.

- When a battery is connected across a semiconductor, the electron experiences a force towards the positive terminal and hole towards the negative terminal.

- The net motion of flow of these charges is called **drift**. This flow of charges constitutes current.

- Electron current is in the direction of flow of electrons. Conventional current is in the direction of flow of holes.

Effect of temperature on intrinsic semiconductor :

- We have seen at absolute zero temperature, semiconductor behaves like insulator. At room temperature, because of thermal energy, few electron-hole pairs are generated which constitute a small current i.e. it has small conductivity.

- Further, if temperature is increased, more electron-hole pairs are generated i.e. **conductivity increases**.

- **Thus as temperature of semiconductor increases, its conductivity increases**.

- **In other words, as the temperature of the semiconductor increases, its resistance decreases**.

- Thus, semiconductor has negative temperature coefficient of resistance.

7.3.2 Extrinsic Semiconductor   (Nov. 18)

- When semiconductors are to be used for specific application, their current carrying capacity must be larger. This is achieved by adding a small amount of suitable impurity to a semiconductor.

- **The process of adding an impurity to a semiconductor is known as doping**.

- The purpose of adding impurity is to increase either the number of free electrons or holes in semiconductors. If a pentavalent impurity (having five electrons in its outermost shell) is added to the semiconductor, a large number of free electrons are produced in the semiconductor.

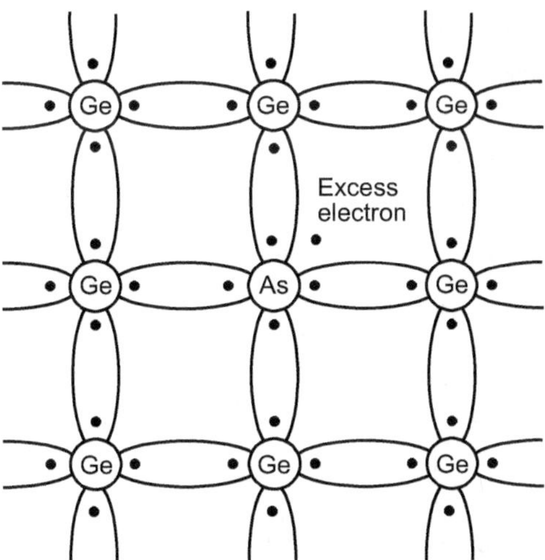

Fig. 7.10 : N-type extrinsic semiconductor

- On the other hand, addition of trivalent impurity (having three electrons) creates a large number of holes in the semiconductor crystal.

- Depending upon the type of impurity added, extrinsic semiconductors are classified into N-type and P-type semiconductors.

- We know that each Si atom consists of 14 electrons and Ge atom consists of 32 electrons. Similarly, Si - 14, Ge - 32, Ar - 33, Sb - 51, P - 15, Ga - 31, In – 49, B - 5.

(a) N-type extrinsic semiconductor :

- When a small amount of pentavalent (five valence electrons) impurity is added to a pure semiconductor, it is known as **N-type semiconductor**. Some **pentavalent impurities are arsenic, antimony, phosphorus, etc**.

- These type of impurities which produce N-type semiconductor are known as donor impurities because they provide free electrons to the crystal semiconductor (Refer Fig. 7.10).

- To understand the formation of an N-type semiconductor, consider a pure Germanium semiconductor crystal. It has four valence electrons which form the covalent bonds. If an impurity like arsenic having five valence electrons (pentavalent impurity) is added to Ge crystal, then arsenic atom replaces one of the Ge atom and four electrons from each Ge and As form the covalent bonds as shown in Fig. 7.10.

- So out of five valence electrons of one impurity atom, one electron remains as a free electron. Even if each atom of impurity provides one free electron, yet an extremely small amount of arsenic impurity provides enough atoms to supply millions of free electrons for current conduction.

- The current conduction in an N-type semiconductor is predominantly electrons (negative charge). So they are called as majority carriers and holes are called as minority carriers.

(b) P-type extrinsic semiconductor :

- When a small amount of trivalent impurity is added to a pure semiconductor, it is called as **P-type semiconductor**.

- The addition of trivalent impurity provides a large number of holes in the semiconductor. Some trivalent impurities are **gallium, indium, boron, aluminium**, etc. The impurities which produce P-type semiconductor are known as **acceptor impurities**.

- To understand the formation of P-type semiconductor, consider a pure Germanium crystal. It has four valence electrons which form the covalent bonds. If an impurity like 'Gallium' is added, then gallium atom replaces one of the Ge atom and three valence electrons from each 'Ge' and 'Ga' form covalent bonds as shown in Fig. 7.11. So out of four electrons of 'Ge', only three have formed the bonds and fourth one cannot form the bond, because 'Ga' atom cannot share any fourth electron. Hence, instead of bond formation, we find a hole as shown.

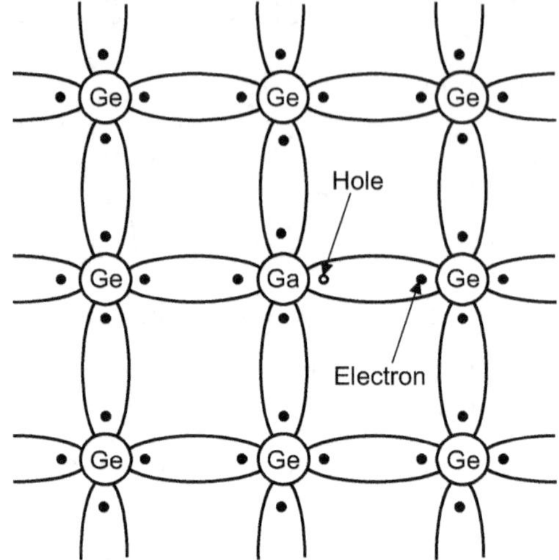

Fig. 7.11 : P-type extrinsic semiconductor

- The current conduction in P-type semiconductor is predominantly by the holes (positive charge). So they are called as **majority carriers** and electrons are called as **minority carriers**.

7.4 PN JUNCTION DIODE - FORWARD AND REVERSE BIASED CHARACTERISTICS   (Nov. 18)

PN junction diode :

- It is a semiconductor device used in most of the electronic circuits. Doped regions meet to form PN junction diode.

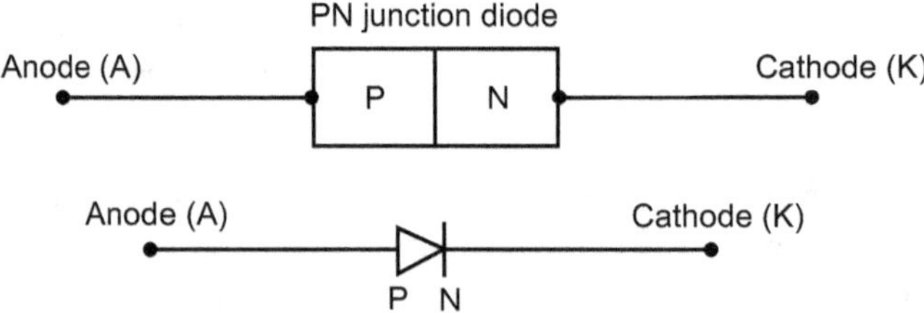

Fig. 7.12 : Schematic symbol or Symbol used in electronic circuit

- In the P region, concentration of holes (+ve charge) is more and in the N region, concentration of free electrons (e⁻) is more.

- When PN junction is formed, some electrons from N region, cross the junction and diffuse into the P region.

- Similarly, some holes from P region, cross the junction and diffuse into the N region. Recombination of holes and electrons takes place in the junction region.

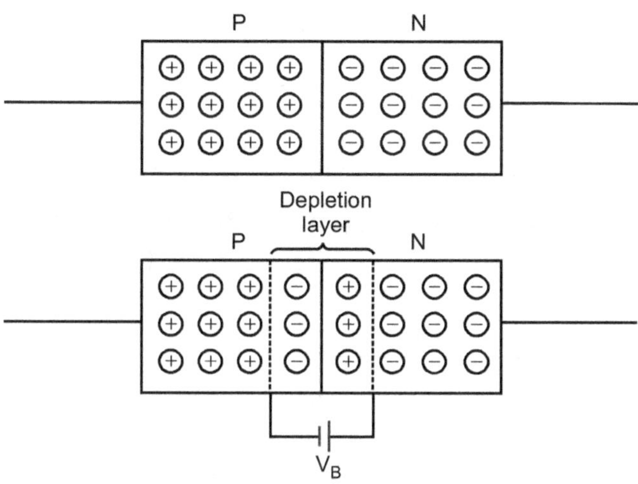

Fig. 7.13 : Formation of depletion layer

- Because of accumulation of holes (+ve) and electrons (–ve) in the depletion layer, an electric field V_B is developed in the depletion layer called barrier potential.

- This electric field V_B acts like a small battery and opposes further diffusion of holes and electrons.

- Barrier potential V_B for silicon (Si) is 0.7 V and barrier potential V_B for Germanium (Ge) is 0.3 V.

Forward biased PN junction diode (Characteristics) :

- External voltage source (battery) is connected across the PN junction diode. If positive terminal of a battery is connected to the 'P' side and negative terminal of the battery is connected to the 'N' side, then diode is said to be **forward biased**.

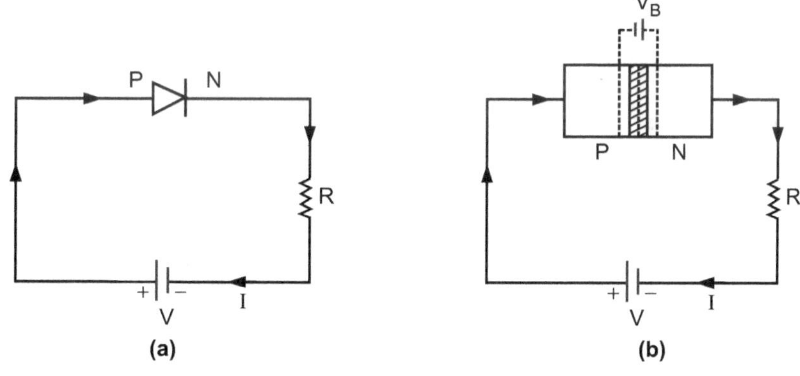

Fig. 7.14

- Holes in the 'P' region get repelled by +ve of the battery and cross the junction, while electrons in the 'N' region get repelled by –ve of battery and cross the junction because of which diode conducts current through it. Diode is conducting.

- Forward biasing of a diode allows easy current flow through the diode and diode acts like a conductor.

- The barrier potential 'V_B' opposes the external battery. In order to flow current through a diode, the external voltage supplied must be greater than V_B (i.e. 0.7 V for Silicon, 0.3 V for Germanium).

- Because of pushing of charge carriers towards the junction in the forward biased diode, the width of the depletion layer decreases.

- If external voltage is increased from zero onwards, initially the forward voltage is increased and values of currents are recorded and graph (i.e. characteristics) is plotted.

- When external voltage is less than 0.6 V, very little diode current flows through the circuit. However, the diode current increases sharply beyond 0.6 V of external forward biased voltage. It is observed that forward voltage drop V_F is nearly the same (constant) as I_F increases.

- Barrier potential for Silicon diode is 0.7 V, and barrier potential for Germanium diode is 0.3 V.

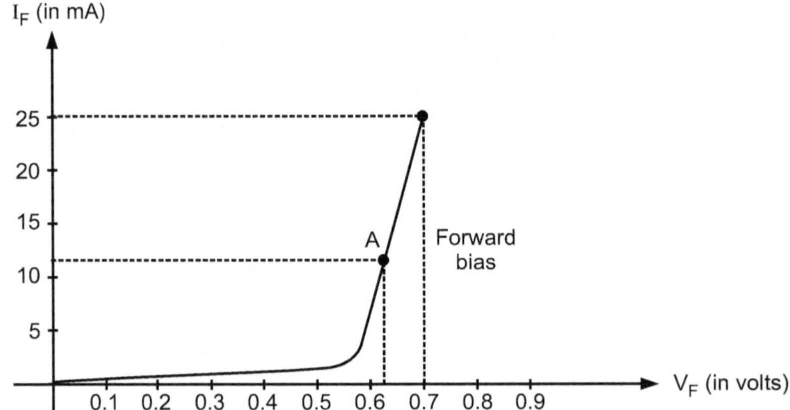

Fig. 7.15 : Forward bias (voltage-ampere i.e. V-I) characteristics of a silicon diode

Reverse biased PN junction diode (Characteristics) :

- External voltage source (battery) is connected across the PN junction diode. If –ve terminal of battery is connected to 'P' side and +ve terminal of battery is connected to 'N' side, then the diode is said to be connected in **reverse bias**.

- Holes in the 'P' region get attracted by –ve of battery and electrons in the 'N' region get attracted by +ve of a battery because of which charge carriers go away from the junction. The effect of such connection is that, charge carriers in both the regions get pulled away from the junction. Because of this pulling away of charge carriers from the junction, the width of the depletion layer increases. The barrier potential V_B assists the external battery.

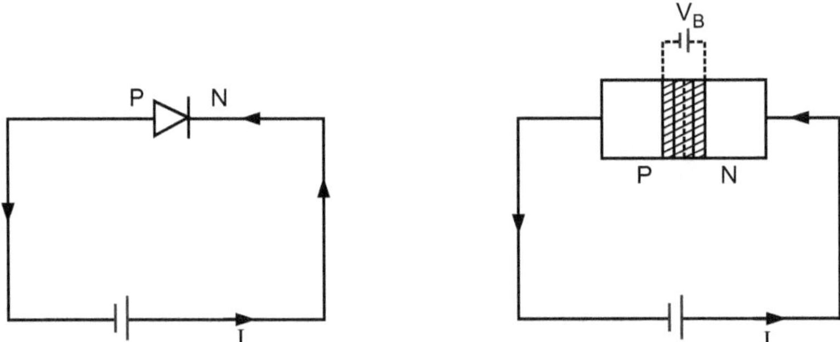

Fig. 7.16 : Reversed biased PN junction diode

- Thus, reverse bias repells the majority carriers away from the junction i.e. they cannot cross the junction and majority current carriers do not flow through the circuit.

Leakage current :

- Though diode does not conduct the majority current carriers, very small amount of current flows through the reverse biased diode, called **leakage current**.

- The leakage current is due to flow of minority carriers in both the regions. i.e. minor carriers in the depletion layer cross the junction. This flow of minority carriers is because of thermal energy.

- The minority carriers mean holes in the N side and electrons in the P side. Any increase in temperature of diode, increases the leakage current.

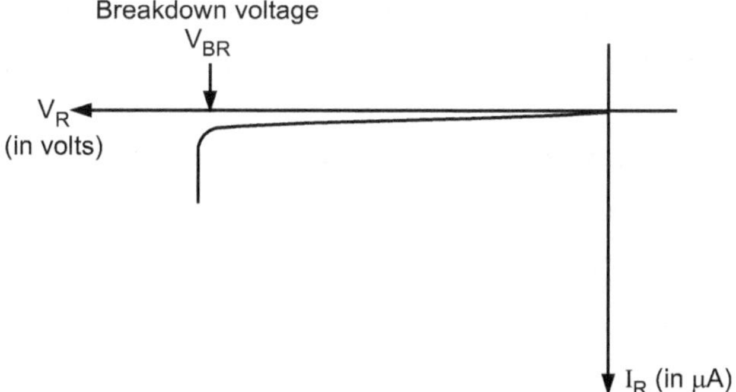

Fig. 7.17 : Reverse bias (voltage-ampere V-I) characteristics of a silicon diode

- Reverse biased voltage is increased and values of current are recorded, and the graph (i.e. characteristics) is plotted.
- The current which flows before breakdown V_{BR} is because of thermally produced minority current carriers. This current is called leakage current which is in microamperes μA.
- As the reverse biased voltage is increased, at a critical voltage V_{BR}, the reverse current through the diode increases sharply. Most of the diodes have breakdown voltage more than 50 V.

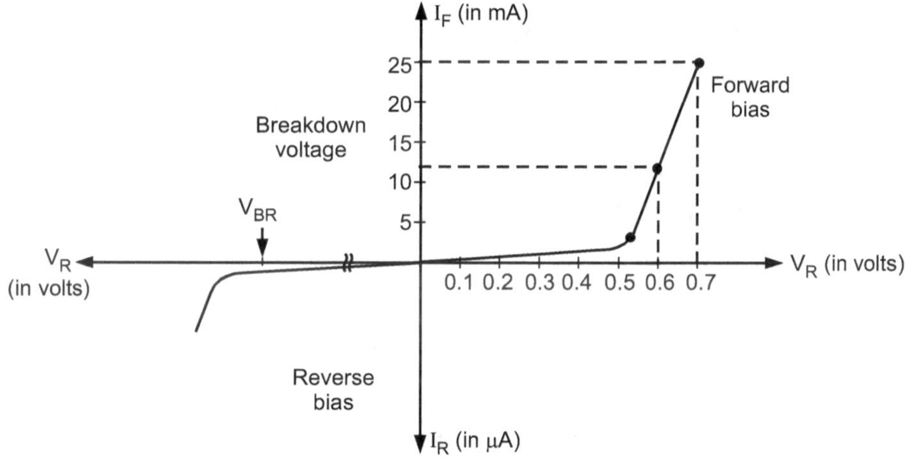

Fig. 7.18 : V-I characteristics of a silicon diode

SUMMARY

- The range of energies possessed by valence electrons is known as valence band.
- The range of energies possessed by conducting electrons is known as conduction band.
- The gap between conduction and valence bands is called as **forbidden gap**.
- In case of conductor, there is overlapping of conduction band and valence band.
- In case of semiconductor, there is gap i.e. forbidden gap of approximately 1 eV.
- In case of insulator, there is wide forbidden gap which is greater than 5.5 eV.
- In intrinsic semiconductor, Ge or Si is taken in pure form.
- In case of N-type extrinsic semiconductor, pentavalent impurity like arsenic, antimony is doped (added) in pure Ge or Si.
- In case of P-type extrinsic semiconductor, trivalent impurity such as gallium or indium is doped (added) into pure Ge or Si.

EXERCISE

1. Discuss the energy bands in (a) conductor, (b) semiconductor and (c) insulator.

2. Define energy band in solids. Draw energy level diagram for (a) conductor, (b) semiconductor and (c) insulator.

3. What is the difference between N-type semiconductor and P-type semiconductor ?

4. What is intrinsic semiconductor ?

5. What is extrinsic semiconductor ?

6. What is meant by doping of a semiconductor ?

7. Write notes on intrinsic and extrinsic semiconductors.

8. Write a note on N-type semiconductor.

9. Write a note on P-type semiconductor.

10. Write a note on band theory of solids.

11. Explain why discrete energy levels of an isolated atom split into a band of energy when atoms combine together to form a crystal.

12. Explain the difference between conductors, insulators and semiconductors using energy band diagram.

13. Sketch a two-dimensional crystal structure of intrinsic silicon at absolute zero temperature.

14. Why is pentavalent impurity known as donor impurity ?

15. Separate out N-type impurities and P-type impurities from the following :

   (a) Boron, (b) Arsenic, (c) Aluminium, (d) Indium, (e) Phosphorus, (f) Antimony.

16. How are valence, conduction and forbidden bands explained ?

17. Distinguish between conductor, semiconductor and insulator.

18. What is the effect of temperature on conductivity of semiconductor ?

19. When silicon doped with boron, is the result N-type or P-type semiconductor ? Give reason.

20. Define P-type and N-type semiconductor.

21. Define conduction band and forbidden energy gap.

22. Name two semiconductors which are commonly used.

23. Classify solids into conductor, semiconductor and insulator on the basis of band theory of solids.

24. Explain covalent bonding in intrinsic and extrinsic semiconductors.

25. Show energy band diagram for conductor and semiconductor.

26. Show diagrammatic representation of P-type and N-type extrinsic semiconductor.

27. What is PN junction diode ?

28. Explain forward biased PN junction diode characteristics.

29. Explain reversed biased PN junction diode characteristics.

30. Draw circuit diagram of forward biased PN junction diode.

31. Draw circuit diagram of reverse biase PN junction diode.

32. Sketch PN junction diode (V-I) characteristics of forward and reverse biased both.

❑❑❑

8

CHAPTER

PHOTOELECTRICITY

8.1   Photoelectric

8.2   Planck's Hypothesis (Planck's Quantum Theory)

8.3   Concept of Photon

8.4   Properties of Photons (Characteristics)

8.5   Characteristics of Photoelectric Effect

8.6   Definitions

8.7   Einstein's Photoelectric Equation

8.8   Photoelectric Cell

8.9   Basic Concept of Solar Energy

Summary

Important Formulae

Solved Examples

Exercise

Problems for Practice

8.1 PHOTOELECTRIC EFFECT   (Nov. 18)

It is found that certain metals like magnesium, zinc, lithium when exposed to ultraviolet light, then electrons are emitted. Some alkali metals like sodium, potassium are sensitive to even visible light. i.e. when light falls on alkali metals, electrons are emitted. Thus, when light of suitable frequency (wavelength) is incident on metallic surface, electrons are emitted from metal surface. As the effect takes place under the influence of light (photo), it is called as **photoelectric effect** and emitted electrons are called as **photoelectrons**.

$$\text{Light energy (photo)} \xrightarrow[\text{into}]{\text{Converted}} \text{Electrical energy}$$

For most of the metals, the photoelectric effect is observed when ultraviolet light falls on them. However, for alkali metals like Na and K, the effect is observed even with visible light. The photoelectric effect was detected by Hertz in 1887.

8.2 PLANCK'S HYPOTHESIS (PLANCK'S QUANTUM THEORY)

Planck proposed the quantum theory for explanation of energy distribution in a black body radiation. According to this theory, energy is not emitted and absorbed continuously, but in a discrete (intermittent) units or packets (bundle). These energy packets are called as **photons** or **quanta**.

The photons are electrically neutral and travel with speed of light i.e. radiation is considered as shower of photons.

If υ is the frequency of light (photon), the energy E associated with photon is directly proportional to υ.

$$\therefore \qquad E \propto \upsilon$$

$$E = \text{Constant} \times \upsilon$$

$$\therefore \qquad \boxed{E = h\upsilon}$$

where, $\qquad$ h = Planck's constant = 6.63×10^{-34} Js

According to this theory, energy is always emitted or absorbed in integral multiple of $h\upsilon$ and not in fraction of $h\upsilon$.

$$\therefore \qquad \boxed{E = n\,h\upsilon} \qquad\qquad \text{where} \quad n = \text{integer} = 1, 2, 3, \ldots$$

8.3 CONCEPT OF PHOTON

According to Planck's theory, the quantum of light wave of frequency 'υ' has the energy

$$E = h\upsilon \qquad \ldots \text{Photon energy}$$

where $\qquad$ h = Planck's constant

$$= 6.63 \times 10^{-34} \text{ J·s}$$

$$= 4.14 \times 10^{-15} \text{ eV·s}$$

Thus light wave has minimum energy = $h\upsilon$ and if it has more energy, the energy of light wave is integral multiple of $h\upsilon$.

Einstein proposed that when light falls on metal (atom), the energy E = $h\upsilon$ of photon is absorbed by the atom i.e. the energy is transferred from light to atom. This is called as absorption of energy.

When light of frequency υ is emitted by an atom, an energy $h\upsilon$ is transferred from atom to light i.e. emission event (photon appears).

Thus photon absorption and photon emission take place.

8.4 PROPERTIES OF PHOTONS (CHARACTERISTICS)

1. **The existence of photon :** The fact or state of existing. The photon is a indivisible entity. The existence of photon is same as existence of electron.

2. **Non-electrical nature of photon :** Photons are electrically neutral. i.e. photons cannot be deflected by electric field.

3. **Non-magnetic nature of photon :** Photons cannot be deflected by magnetic field.

4. Photons travel with speed of light i.e. 3×10^8 m/s.

5. Photons do not ionize.

6. **Mass and momentum of a photon :** Mass and energy are equivalent (Einstein's theory of relativity). The mass of photon is given by

$$m = \frac{E}{c^2} = \frac{h\upsilon}{c^2} = \frac{h}{c\lambda}$$

where
$$m = \text{Mass of photon}$$
$$E = \text{Energy of photon}$$
$$c = \text{Speed of light}$$
$$\upsilon = \text{Frequency of radiation}$$
$$\lambda = \text{Wavelength of radiation}$$
$$h = \text{Planck's constant}$$

and the momentum of photon is given by

$$\text{Momentum} = \text{Mass} \times \text{Velocity}$$

$$= m \times c = \frac{h\upsilon}{c^2} \times c$$

$$= \frac{h\upsilon}{c} = \frac{h}{\lambda}$$

8.5 CHARACTERISTICS OF PHOTOELECTRIC EFFECT

(1) A metal emits electrons only when the incident (light) radiation has frequency greater than critical frequency (υ_0) called threshold frequency. Threshold frequency υ_0 is different for different metals.

(2) Photoelectric current is directly proportional to intensity of light and is independent of frequency.

(3) The velocity of photoelectron is directly proportional to frequency of light and is independent of intensity.

(4) For a given metal surface, stopping potential is directly proportional to frequency and is not dependent on intensity of incident light.

(5) The rate of emission of photoelectrons from the photocathode is independent of its temperature i.e. photoelectric emission is different from thermionic emission.

(6) This process is **instantaneous** i.e. the emission of photoelectrons starts at the moment - light is incident on the metal surface.

8.6 DEFINITIONS   (Nov. 18)

- **Threshold frequency (υ_0) :** Threshold frequency υ_0 of metal is the minimum frequency of the incident light at which emission just begins. υ_0 changes from metal to metal.

- **Threshold wavelength (λ_0) :** Threshold wavelength λ_0 of metal is the maximum wavelength of light at which emission just begins.

- **Photoelectric work function (W_0) :** Photoelectric work function W_0 of metal is the energy required to detach the electron from metal.

- **Stopping potential :** Stopping potential of photoelectric cell is the negative potential given to cell at which photoelectric current becomes zero.

8.7 EINSTEIN'S PHOTOELECTRIC EQUATION

According to quantum theory, radiation is considered as shower of particles called photons. Each photon carries energy $E = h\upsilon$. These photons travel with speed of light. When radiation falls on the metal, these photons collide with the metal atoms. During collision, energy of the photon is absorbed by the metal atom. Thus, now energy of an atom is $h\upsilon$. Part of this energy is used to detach (loose) the electron from the atom and the remaining energy is used in giving kinetic energy to the electron.

Energy of photon absorbed by the atom ($h\upsilon$) is

(i)   used to detach the electron (W_o) and

(ii)   given to electron in the form of kinetic energy.

Thus, $h\upsilon = W_o + K.E.$

$$h\upsilon = W_o + \frac{1}{2} mv^2$$

$\therefore$ $\boxed{\frac{1}{2} mv^2 = h\upsilon - W_o}$

where W_o = photoelectric work function

$$W_o = h\upsilon_o$$

$\therefore$ $\frac{1}{2} mv^2 = h\upsilon - h\upsilon_o$

$\boxed{\frac{1}{2} mv^2 = h(\upsilon - \upsilon_o)}$

This equation is Einstein's photoelectric equation,

where, m = Mass of electron

v = Velocity of electron

h = Planck's constant

υ = Frequency of radiation

υ_o = Threshold frequency

Cases :

(i)   If $\upsilon < \upsilon_o$ ...... K.E. is negative ...... meaningless ...... no emission.

(ii)   If $\upsilon = \upsilon_o$ ...... K.E. is zero ...... emission just begins.

(iii) If $\upsilon > \upsilon_o$ ...... K.E. is positive ...... emission takes place.

As υ increases, $(\upsilon - \upsilon_o)$ also increases, therefore $\frac{1}{2} mv^2$ i.e. K.E. increases (velocity increases).

$\therefore$ $K.E. \propto$ Frequency υ

8.8 PHOTOELECTRIC CELL

Principle : Light energy is converted into electrical energy.

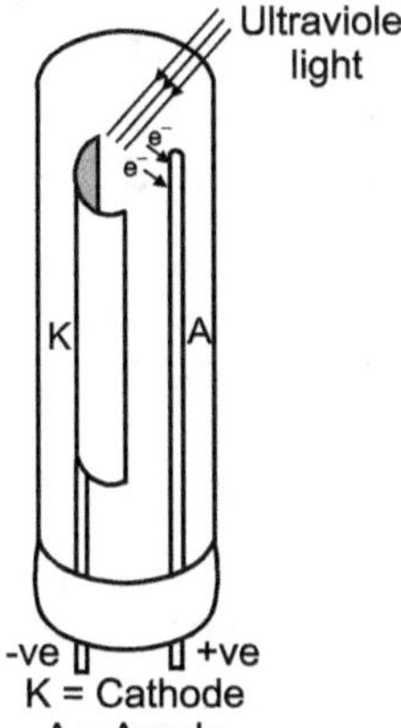

Fig. 8.1

Photoelectric cell consists of (i) cathode K and (ii) anode A, enclosed in an evacuated glass bulb.

The semi-cylindrical cathode coated with photosensitive material forms inner side. The anode is a rod of platinum kept along the axis of cathode. The cathode is connected to negative terminal and anode is connected to positive terminal of high tension battery through milliameter.

When light is allowed to fall on cathode, it emits photoelectrons. These electrons are attracted by the anode.

The photoelectric current flows through the circuit and milliammeter shows a deflection.

Types of Photoelectric Cell :

There are three types of photoelectric cell.

1. Photoemissive cell : When light of particular frequency is allowed to fall on a cathode of a photo cell, electrons are emitted. This type of cell requires external battery supply.

2. Photoconductive cell : In case of photoconductive cell, when light falls on a cell, its resistance decreases i.e. conductivity increases. It also requires external battery supply.

3. Photovoltaic cell : In this type of cell, light energy falling on it is directly converted into electrical energy. It does not require external battery supply.

Applications of Photoelectric Cell

1. Photoelectric cell is used in lux-meter to measure the intensity of light.
   (**Principle** - Photoelectric current is directly proportional to intensity of light.)

2. It is used for automatic control of traffic signals.

3. It is used to switch on and off automatically the street lights.

4. It is used in recording and reproduction of sound during shooting of a film.

5. Photoelectric cells are used in television sets, fire alarms.

6. It is used in detecting flaws in metals.

7. Photoelectric cell is used in **Burglar alarm.**

Burglar alarm : It is the alarm used to protect valuables (money and gold ornaments) from thieves. The infrared light from the source is allowed to fall on photoelectric cell. Photoelectric cell is in ON condition and photoelectric current flows continuously in the circuit. Photoelectric cell is kept at the entry or near valuables. When some one come near the valuables, infrared light gets interrupted and photoelectric current immediately stops which automatically starts the electric alarm (bell) and thus entry of thieves is detected.

8. Photoelectric cell is used in **exposure meter.**

Exposure meter : Exposure meter is a device which is used to determine exact time for which the lens of camera must be exposed to light for good photograph. We know that photoelectric current is directly proportional to intensity of light. In bright light, the photoelectric current will be high and lens of camera must be exposed for shorter time. The milliameter scale in exposure meter is calibrated to read the time directly.

Similarly in poor light, photoelectric current is small and lens of camera must be exposed for longer time. Direct time can be read in meter.

8.9 BASIC CONCEPT OF SOLAR ENERGY

8.9.1 Solar Energy

The energy that we receive from the sun is called as the solar energy. The energy produced by the sun is the most important source of energy for all life forms on the earth. It is renewable source of energy. Solar energy means capturing the rays of sun and storing its heat. This heat can be converted with the help of solar panels into heat or electrical energy.

Solar technologies use the solar energy to light homes and streets, to heat the water, to heat the homes and also to produce electricity. The main benefit of solar energy is that it does not produce any pollutants and is one of the cleanest source of energy. Solar energy helps in lessening the green-house effect. Therefore, it is ecologically acceptable source of energy.

8.9.2 Importance of Solar Energy

Solar energy is every important for survival of life on the earth. Human beings, plants, animals all require solar energy. Plants require solar energy to produce oxygen and to prepare food (i.e. the process of photosynthesis). Solar energy is also required to produce both pure and salt water in oceans as it is the only source of melting frozen ice formed on the mountain caps. Apart from that, solar energy is also used to produce electricity which can be used to run various machines.

The fossil fuels and other gases and oil that are extracted from the mines are non-renewable energy sources. Also they are costly and cause lot of pollution. But solar energy is renewable source of energy which can be used for lots of activities. Also it is available free of cost. As fossil fuels and other oils are soon going to disappear, the solar energy which is abundant is very important and it should be utilized well.

8.9.3 Development of Solar Energy in the World

Solar energy has been used since prehistoric times, but in a most primitive manner. Before 1970, some research and development was carried out in a few countries to exploit solar energy more efficiently, but most of this work remained mainly academic. The oil crisis of 1970 revealed the delicate nature of fossil fuels as a source of energy for the world. After the dramatic rise in oil prices in 1970s, several countries began to formulate extensive research and development programs to exploit solar energy.

Solar energy is developed in U.S.A., Japan, France, Israel, Australia and India. About 80% of the solar energy is produced by U.S.A., particularly in the deserts of California. The largest solar power plant in the world is located in the Mojave Desert in California, covering 1000 acres. Now Sri Lanka and Indonesia also have planned to use the solar energy for electrification in the rural areas. In the next few years, it is expected that millions of households in the world will be using solar energy as the trends in USA and Japan show. In India also, the Indian Renewable Energy Development Agency and the Ministry of Non-conventional Energy Sources are formulating a program to have solar energy in more than a million households in the next few years. However, people's initiative is essential to make the program successful.

8.9.4 Development of Solar Energy in India

India has tropical location and it is one the few countries with long days and plenty of sunshine. Therefore, there is tremendous potential for the development of solar energy in India. The areas with abundent sunshine are suitable for harnessing solar energy for a number of applications. Solar thermal energy is being used in India for heating water for both domestic and industrial purposes.

At present, solar photovoltaic (SPV) systems in India are used in remote villages, islands, hill tops for lighting and communication, TV power supply, water pumping for agriculture, radios, refrigerators, etc. The SPV systems are also used as power plant for meeting electricity needs of voltages, schools, public health centres, etc. In Delhi, Haryana and Himachal Pradesh, small pumps operated by solar power are being used for irrigation and drinking water supply. In many remote rural areas of West Bengal, due to Indian government policy to provide subsidy, the solar energy is used in the forms of solar lanterns, solar pumps, solar streetlights, etc. Under the program for the electrification of remote villages through SPV system, thousands of villages have been electrified. The world's largest solar cooking system, for cooking food for 15000 to 20000 people has been installed in October 2002, at Tirupati Devasthan in Andhra Pradesh, and it is still functioning well. A 140 MW integrated solar power plant is to be set up in Jodhpur, but the initial expense incurved is still high. With new advancements in scientific researches, solar energy could be more affordable in future with decreasing costs and increasing efficiency.

8.9.5 Uses of Solar Energy

Solar energy in the form of thermal energy can be used for cooking/heating, Drying/Timber seasoning, Distillation, Electricity/Power generation, Cooling, Refrigeration, Cold storage, etc. Some of the gadegets and devices working on thermal energy are Solar cookers, Concentrating collectors, Solar hot water systems (Domestic and Industrial), Solar hot air systems, Solar driers, Solar timber kilns, Solar photovoltaic systems, etc. Solar powered hot water systems utilize solar energy to heat water. Solar chimneys are passive solar ventilation systems. Solar energy can be used for making potable, brackish or saline water. Food can be cooked, dried or pasteurized using solar energy. Space missions by various countries use solar energy to power their spaceships. Solar energy is the preferred mode of creating power where the need is temporary, e.g. temporary fairs, mining sites, olympics, etc. Solar driers are used to dry vegetables, fish, grapes, tea, tobacco, etc. Solar energy can help in producing power for watches and calculators that do not work on batteries.

Solar energy can also be used to meet our electricity requirements. Solar radiations are converted directly into DC electricity through solar photovoltaic (SPV) cells. This electricity can either be used as it is or it can be stored in the battery. This stored electricity can be used at night. SPV systems can be used for a number of applications such as Domestic lighting, Street lighting, Village electrification, Water pumping, Desalination of salty water, Powering of remote telecommunication repeater stations and Railway signals.

The efforts are being made to make the efficient use of solar energy. This would reduce our dependence on non-renewable sources of energy and make our environment cleaner.

8.9.6 Advantages of Solar Energy

(1) Solar energy is the most readily available source of energy.

(2) Solar energy system can work independently without any connection. So it can be utilized and installed in remote areas like forests, deserts, mountains, etc. where there is no sign of electricity.

(3) It helps in decreasing the harmful gases and does not contribute to acid rains, global warming, forest destruction and other natural disasters.

(4) It is very clean and environment friendly source of energy and do not cause pollution.

(5) It is non-exhaustible source of energy.

8.9.7 Disadvantages of Solar Energy

(1) Solar energy is available only during daytime with clear sky and bright sunshine. So for night use, storage battery is required, which is very costly.

(2) Generation of solar energy is restricted in the areas with cloudy climate.

(3) The solar collectors, solar panels and solar cells that are used to absorb heat from the sun are very expensive.

(4) The solar batteries that are to be charged are very heavy and require large storage space. Replacing it is also difficult.

(5) Installation of solar energy plants requires large area so that system can provide good amount of electricity. This is the great disadvantage in places where the area is small.

8.9.8 Impacts of solar energy on Environment :

Solar energy does no produce air or water pollution or greenhouse gases. However, using solar energy may have some indirect negative impacts on the environment.

For example :

(1) To make the photovoltaic (PV) cells, that convert sunlight into electricity, some toxic materials and chemicals are used.

(2) Some solar thermal systems use potentially hazardous fluids to transfer heat.

(3) Large solar power plants can affect the environment neat their locations.

(4) Some power plants require water for cleaning solar collectors and concentrators or for cooling turbine generators. Using large amount of ground water or surface water in some arid locations may affect the ecosystems that depends on these water resources.

(5) Clearing land for construction and placement of the power plant may have long-term effects on habitat areas for native plants and animals.

(6) Also the beam of sunlight created by solar power tower can kill birds and insects that fly into the beam.

SUMMARY

- When light of frequency more than threshold value, falls on a metal, then electrons are ejected out. Thus, light energy is converted into electrical energy.

$$\text{Light energy (photo)} \xrightarrow[\text{into}]{\text{converted}} \text{Electrical energy.}$$

  This is photoelectric effect.

- According to Planck's theory, light is emitted in the form of energy packets called photons or quanta.

- Energy possessed by photon or quanta is

$$E = h\upsilon$$

  where υ – frequency of light

- Emission takes place only if frequency of light is greater than threshold frequency.

- Photoelectric current $\propto$ intensity of incident light.

- Velocity of photoelectron $\propto$ frequency of incident light.

- Threshold frequency υ_0 of metal is the minimum frequency of incident light at which emission just begins.

- Photoelectric cell converts light energy into electrical energy.

   When light falls on photo cell, it becomes ON.

- Photomultiplier tube is a device which increases the value of photoelectric current.

- $\frac{1}{2} mv^2 = h(\upsilon - \upsilon_0) \ldots$ Einstein's equation.

- When light falls on photo cell, it becomes ON and when intensity of light increases then photoelectric current increases. This characteristic plays major role in number of applications.

- Solar energy means capturing the sun rays and storing its heat. This heat can be converted into heat or electrical energy with the help of solar panels.

- Solar energy is renewable energy source which does not produce air or water pollution or greenhouse gases.

IMPORTANT FORMULAE

1. $E = h\upsilon = \dfrac{hc}{\lambda}$, where
    - E - Energy of photon
    - h - Planck's constant
    - υ - Frequency of radiation

2. $E = n\,h\upsilon = \dfrac{nhc}{\lambda}$
    - c - Speed of light
    - λ - Wavelength of light
    - n - Number of photons

3. $\frac{1}{2} mv^2 = h(\upsilon - \upsilon_0) = hc\left(\dfrac{1}{\lambda} - \dfrac{1}{\lambda_0}\right)$, where
    - m - Mass of photo-electron
    - v - Velocity
    - υ - Frequency of radiation
    - υ_0 - Threshold frequency
    - λ - Wavelength
    - λ_0 - Threshold wavelength

4. $W_0 = h\upsilon_0 = \dfrac{hc}{\lambda_0}$, where W_0 - Photoelectric work function

    or energy required to loose the electron.

5. Conversions :
$$1 \text{ eV} = 1.6 \times 10^{-19} \text{ joules}$$
$$1 \text{ A.U. or } 1 \text{ A}^\circ = 1 \times 10^{-10} \text{ m}$$

6. Data : Take speed of light,
$$c = 3 \times 10^8 \text{ m/s}$$
$$h = 6.62 \times 10^{-34} \text{ J-s if not given.}$$

SOLVED EXAMPLES

Example 8.1 : *An accelerated electron emits a quantum of radiation with a frequency 8×10^{18} cycles per second. Calculate energy of electron.*

Planck's constant, $h = 6.62 \times 10^{-34}$ joule-sec.

Solution : $E = h\upsilon = 6.62 \times 10^{-34} \times 8 \times 10^{18}$

$\therefore$ $\boxed{E = 5.296 \times 10^{-15} \text{ joule}}$

Example 8.2 : *The photoelectric work function of certain metal is 3×10^{-19} joules. Calculate its threshold frequency. Planck's constant is 6.625×10^{-34} Js.*

Solution : Given : $W_0 = 3 \times 10^{-19}$ J

$h = 6.625 \times 10^{-34}$ Js

$\upsilon_0 = ?$

We have, $W_0 = h\upsilon_0$

$\therefore$ $\upsilon_0 = \dfrac{W_0}{h} = \dfrac{3 \times 10^{-19}}{6.625 \times 10^{-34}}$

$\boxed{\upsilon_0 = 0.453 \times 10^{15} \text{ Hz}}$

Example 8.3 : *The photoelectric work function of a metal is 5 eV. Calculate the threshold frequency and threshold wavelength. $h = 6.6 \times 10^{-34}$ Js, $c = 3 \times 10^8$ m/s.*

Solution : Given : $W_0 = 5$ eV

$= 5 \times (1.6 \times 10^{-19})$ J

$\upsilon_0 = ?$

We have, $W_0 = h\upsilon_0$

$\therefore$ $\upsilon_0 = \dfrac{W_0}{h} = \dfrac{(5 \times 1.6 \times 10^{-19})}{(6.6 \times 10^{-34})}$

$\boxed{\upsilon_0 = 1.212 \times 10^{15} \text{ Hz}}$

and $\upsilon_0 = \dfrac{c}{\lambda_0}$

$\therefore$ $\lambda_0 = \dfrac{c}{\upsilon_0} = \dfrac{3 \times 10^8}{(1.212 \times 10^{15})}$

$\lambda_0 = 2.475 \times 10^{-7}$ m

$\boxed{\lambda_0 = 2475 \times 10^{-10} \text{ m}}$

or $\boxed{\lambda_0 = 2475 \text{ A}°}$

Example 8.4 : *The threshold frequency of a metal is 1.2×10^{15} Hz. If a light of frequency 1.5×10^{15} Hz is made incident on the metal plate, calculate the maximum K.E. of the ejected photoelectron. (Take $h = 6.62 \times 10^{-34}$ Js)*

Solution : Given : $\upsilon_0 = 1.2 \times 10^{15}$ Hz

$\upsilon = 1.5 \times 10^{15}$ Hz

$$KE_{max} = ?$$

$$h = 6.62 \times 10^{-34} \text{ Js}$$

$$KE_{max} = h\,(v - v_0) = 6.62 \times 10^{-34} \times (1.5 \times 10^{15} - 1.2 \times 10^{15})$$

$$\boxed{KE_{max} = 1.986 \times 10^{-19} \text{ J}}$$

Example 8.5 : *The threshold frequency of metal is 1.12×10^{15} Hz. If a light of frequency 1.39×10^{15} Hz is made incident on the metal plate, calculate the maximum kinetic energy of the ejected photoelectrons.*

Solution : Given : $v_0 = 1.12 \times 10^{15}$ Hz, $\quad v = 1.39 \times 10^{15}$ Hz

$$K.E. = ?, \quad h = 6.63 \times 10^{-34} \text{ Js}$$

We know, $\quad K.E. = h\,(v - v_0)$

$$= 6.63 \times 10^{-34} \times (1.39 - 1.12) \times 10^{15} = 6.63 \times 0.27 \times 10^{-19}$$

$\therefore \qquad \boxed{K.E. = 1.79 \times 10^{-19} \text{ J}}$

Example 8.6 : *The threshold wavelength for silver is 3800 A°. Calculate maximum energy in eV of photoelectrons emitted when U.V. light of wavelength 2600 A° is incident on it.*

Solution : Given : $\quad \lambda_0 = 3800 \text{ A°} = 3800 \times 10^{-10}$ m

$$E = ?$$

$$\lambda = 2600 \text{ A°} = 2600 \times 10^{-10} \text{ m}$$

$$c = 3 \times 10^{8} \text{ m/s (speed of light)}$$

$$E = h\,(v - v_0)$$

but $\qquad v = \dfrac{c}{\lambda}$

$$v_0 = \dfrac{c}{\lambda_0}$$

$$E = h\left(\dfrac{c}{\lambda} - \dfrac{c}{\lambda_0}\right) = hc\left(\dfrac{1}{\lambda} - \dfrac{1}{\lambda_0}\right)$$

$$= 6.63 \times 10^{-34} \times 3 \times 10^{8} \times \left(\dfrac{1}{2600 \times 10^{-10}} - \dfrac{1}{3800 \times 10^{-10}}\right) = 2.415 \times 10^{-19} \text{ J}$$

$$E = \dfrac{2.415 \times 10^{-19}}{1.6 \times 10^{-19}}$$

$$\boxed{E = 1.5 \text{ eV}}$$

Example 8.7 : *Find maximum K.E. of photoelectrons ejected from surface of metal of light of frequency 1×10^{15} Hz. (Given : Threshold wavelength for metal = 4500 A°).*

Solution : $\qquad KE_{max} = ?$

$$v = 1 \times 10^{15} \text{ Hz}$$

$$\lambda_0 = 4500 \text{ A°} = 4500 \times 10^{-10} \text{ m}$$

Assume $\qquad h = 6.625 \times 10^{-34} \text{ Js}$

$$c = 3 \times 10^{8} \text{ m/s}$$

$$K.E. = h\,(v - v_0)$$

$$= h\left(v - \dfrac{c}{\lambda_0}\right) = 6.625 \times 10^{-34} \times \left(1 \times 10^{15} - \dfrac{3 \times 10^{8}}{4500 \times 10^{-10}}\right)$$

$$\boxed{K.E. = 2.21 \times 10^{-19} \text{ J}}$$

Example 8.8 : *The photoelectric work function of a certain metal is 3×10^{-19} joules. Calculate it's threshold frequency if Planck's constant is 6.625×10^{-34} Js.*

Solution : Given : $W_o = 3 \times 10^{-19}$ J, $\quad \upsilon_o = ?, \quad h = 6.625 \times 10^{-34}$ J-s

We have, $\qquad W_o = h \upsilon_o$

$\therefore \qquad \upsilon_o = \dfrac{W_o}{h} = \dfrac{3 \times 10^{-19}}{6.625 \times 10^{-34}}$

$$\boxed{\upsilon_o = 4.528 \times 10^{14} \text{ Hz}}$$

Example 8.9 : *The energy of photon is 5.28×10^{-19} J. Calculate frequency and wavelength. (Given : $c = 3 \times 10^8$ m/s and $h = 6.625 \times 10^{-34}$ J-s)*

Solution : Given :

$$E = 5.28 \times 10^{-19} \text{ J}$$

$$c = 3 \times 10^8 \text{ m/s}$$

$$h = 6.625 \times 10^{-34} \text{ J-s}$$

$$\upsilon = ?$$

$$\lambda = ?$$

We have, $\qquad E = h\upsilon \quad \text{and} \quad E = \dfrac{hc}{\lambda}$

$$E = h\upsilon$$

$\therefore \qquad \upsilon = \dfrac{E}{h} = \dfrac{5.28 \times 10^{-19}}{6.625 \times 10^{-34}}$

$\therefore \qquad \boxed{\upsilon = 0.797 \times 10^{15} \text{ Hz}}$

and $\qquad E = \dfrac{hc}{\lambda}$

$\therefore \qquad \lambda = \dfrac{hc}{E} = \dfrac{(6.625 \times 10^{-34}) \times (3 \times 10^8)}{(5.28 \times 10^{-19})}$

$\qquad \lambda = \dfrac{6.625 \times 3}{5.28} \times 10^{-34 + 8 + 19}$

$\qquad \lambda = 3.764 \times 10^{-4} \text{ m}$

$\therefore \qquad \boxed{\lambda = 3764 \times 10^{-10} \text{ m}}$

or $\qquad \boxed{\lambda = 3764 \text{ A}^\circ}$

Example 8.10 : *An accelerated electron emits a quantum of radiation with frequency 8×10^{18} Hz. Calculate the energy of electron in electron volt. (Given : $h = 6.625 \times 10^{-34}$ Js.)*

Solution : Given : $\quad \upsilon = 8 \times 10^{18}$ Hz

$$h = 6.625 \times 10^{-34} \text{ Js}$$

$$E = ?$$

$$E = h \times \upsilon$$

$$= (6.625 \times 10^{-34}) \times (8 \times 10^{18}) \text{ joule}$$

but $\quad 1 \text{ eV} = 1.6 \times 10^{-19}$ joules

$\therefore \qquad 1 \text{ joule} = \dfrac{1}{(1.6 \times 10^{-19})} \text{ eV}$

$$\therefore \qquad E = \frac{(6.625 \times 10^{-34}) \times (8 \times 10^{18})}{(1.6 \times 10^{-19})} \text{ eV}$$

$$E = 33.125 \times 10^{-34 + 18 + 19}$$

$$\boxed{E = 33.125 \times 10^{3} \text{ eV}}$$

Example 8.11 : *The photoelectric work function for a certain metal is 3.2 eV. Calculate the threshold frequency. (Given : h = 6.63 × 10^{-34} Js.)*

Solution : Given : $W_0 = 3.2$ eV $= 3.2 \times 1.6 \times 10^{-19}$ joules

$$h = 6.63 \times 10^{-34} \text{ Js}$$

$$\upsilon_0 = ?$$

We have, $\qquad 1$ eV $= 1.6 \times 10^{-19}$ joules

$$W_0 = h\upsilon_0$$

$$\therefore \qquad \upsilon_0 = \frac{W_0}{h} = \frac{(3.2 \times 1.6 \times 10^{-19})}{(6.63 \times 10^{-34})}$$

$$\upsilon_0 = 0.772 \times 10^{-19 + 34}$$

$$\therefore \qquad \boxed{\upsilon_0 = 0.772 \times 10^{15} \text{ Hz}}$$

EXERCISE

1. State Planck's hypothesis of quantum radiation.
2. Explain 'photoelectric effect'.
3. State characteristics of photoelectric effect.
4. State Einstein's equation of photoelectric effect.
5. Define 'threshold frequency'.
6. Describe different types of photoelectric cells.
7. State applications of photoelectric cells and explain two of them.
8. State whether photoelectron possess K.E. in following cases or not :
   (i)   If $\upsilon < \upsilon_0$
   (ii) If $\upsilon = \upsilon_0$
   (iii) If $\upsilon > \upsilon_0$

   where υ = frequency of incident radiation
   $\qquad \upsilon_0$ = threshold frequency

9. Explain the use of photoelectric cell in Burglar alarm.
10. State the characteristics of photoelectric effect.
11. Explain Planck's hypothesis.
12. If the energy of light incident on photosensitive material is equal to work function of that material, calculate velocity with which photoelectrons are emitted.
13. State in following cases whether photoelectric effect takes place or not.
   (i)   If $\upsilon > \upsilon_0$.
   (ii) If $\upsilon < \upsilon_0$.

   where, υ = Frequency of incidence
   $\qquad \upsilon_0$ = Threshold frequency

14. State whether following statements are true or false.

    (i)   Electron volt is the unit of energy.

    (ii) Light is emitted from light source in continuous manner.

15. What is solar energy ? Give its importance.

16. Write a note on : (i) Development of solar energy in the world, (ii) Development of solar energy in India.

17. Write the uses of solar energy.

18. What are the advantages of solar energy ?

19. Give any four disadvantages of solar energy.

20. What are the negative impacts of solar energy on environment ?

PROBLEMS FOR PRACTICE

1. The energy of a photoelectron is 3.3 eV. What is its frequency ?   (**Ans.** 7.96×10^{14} Hz)

2. When a light of wavelength 5400 A° is incident on a metal plate, the electrons are emitted with zero velocity. Calculate the threshold frequency and work function.

    (**Ans.** $v_0 = 5.55 \times 10^{14}$ Hz, $W_0 = 3.69 \times 10^{-19}$ J or $W_0 = 2.29$ eV)

3. The photoelectric work function of a photosensitive material is 3×10^{-19} J. Calculate its threshold wavelength.   (**Ans.** $v_0 = 4.52 \times 10^{14}$ Hz, $\lambda_0 = 6630$ A°)

4. The photoelectric work function of a certain metal is 2.92×10^{-10} J. Calculate its threshold frequency.   (**Ans.** 4.42×10^{23} Hz)

5. If a light of wavelength 4000 A° is incident on a metal surface of work function 5 eV, will the electrons be ejected or not ?

    (**Ans.** $v_0 = 1.21 \times 10^{15}$ Hz, $v = 0.75 \times 10^{15}$ Hz, Since $v < v_0$, electrons will not be emitted)

6. The threshold frequency of a metal is 1.11×10^{15} Hz. If a light of frequency 1.49×10^{15} Hz is made incident on the metal plate, calculate the maximum kinetic energy of the ejected photoelectron.   (**Ans.** K.E. $= 2.52 \times 10^{-19}$ joule)

7. The photoelectric work function of a certain metal is 3×10^{-19} joules. Calculate its threshold frequency, if Planck's constant is 6.625×10^{-34} Js.   (**Ans.** $v_0 = 0.45 \times 10^{15}$ Hz)

8. Find the energy of quantum for wavelength 6000 A°, velocity of light $= 3 \times 10^8$ m/s and Planck's constant $= 6.625 \times 10^{-34}$ Js.   (**Ans.** E $= 2.07$ eV or E $= 3.31 \times 10^{-19}$ J)

9. Find the maximum K.E. of photoelectrons ejected from surface of metal of light of frequency 2×10^{15} Hz. (Given : Threshold wavelength for metal $= 5200$ A°).   (**Ans.** K.E. $= 9.43 \times 10^{-19}$ J)

9

CHAPTER

LASER

9.1   Light Amplification by Stimulated Emission of Radiation (LASER)

9.2   Working Principle of Laser

9.3   Absorption or Stimulated Absorption

9.4   Spontaneous Emission

9.5   Stimulated Emission

9.6   Population Inversion

9.7   Pumping Methods

9.8   Optical Pumping (Three Energy Level Laser System)

9.9   Properties or Characteristics of Laser

9.10   Helium-Neon Laser (He-Ne Gas Laser)

9.11   Applications of Laser

       Summary

       Exercise

9.1 LIGHT AMPLIFICATION BY STIMULATED EMISSION OF RADIATION (LASER)

The term LASER stands for Light Amplification by Stimulated Emission of Radiation.

The laser is a device that produces a light beam with some remarkable properties :

(1) The light is coherent : The light with the waves, all exactly in same phase.

(2) The light is monochromatic : The light whose waves all have the same frequency or wavelength.

(3) The light has unidirectionality : The light produces sharp focus.

(4) The beam is extremely intense : The light has extreme brightness.

In order to understand the concept properly observe the pictorial difference given in Fig. 9.1.

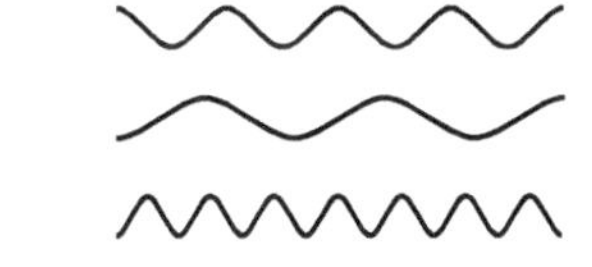

(a) Ordinary light (phase and frequency/wavelength mismatching)   **(b) Monochromatic incoherent light (frequency/wavelength is same but phase is mismatching)**   **(c) Monochromatic coherent light (LASER) (phase, frequency/wavelength all are matching)**

Fig. 9.1

(9.1)

9.2 WORKING PRINCIPLE OF LASER

We know that in order to force water from lower level to higher level, we need some energy and similarly energy is given out when water falls from higher level to lower level.

An atom must be imparted with some energy so that it gets excited from lower energy level to higher energy level. Similarly energy is emitted by an atom when transition takes place from higher energy level to lower energy level. Emission of energy is equal to the difference of two energy levels. Emission is in the form of photon.

i.e. $\qquad E = h\upsilon = (E_2 - E_1)$

where, $\qquad$ E – Energy of photon

$\qquad$ h – Planck's constant

$\qquad$ υ – Frequency of radiation

$\qquad$ E_2 – Energy of higher level

$\qquad$ E_1 – Energy of lower (ground) level

When such a photon is incident on an atom, then atom absorbs this energy and gets excited i.e. moves from lower energy level E_1 to higher energy level E_2 (excitation). [Refer Fig. 9.2 (a)]. This excited state is unstable (non-equilibrium). The atom is not willing to remain in excited state for longer time. In course of time, the atom jumps from excited state to lower state. During this transition, emission of energy takes place in the form of photon.

The atom can remain for unlimited time in ground state but it can remain in excited state for limited time only. This limited time for which atom remains in excited state is known as life time. The life time of excited hydrogen atom is of the order of 10^{-8} sec.

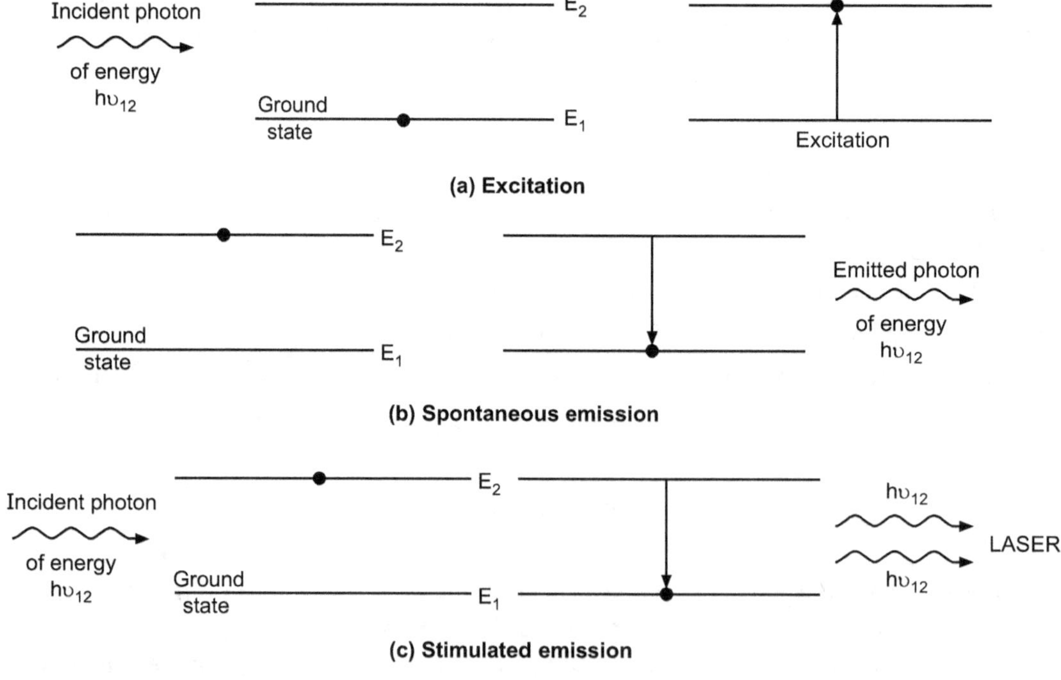

Fig. 9.2

Soon after the life time of excited state, atom returns to ground state (de-excitation) by emitting a photon of energy $h\upsilon$. This emission process is known as spontaneous emission. [Refer Fig. 9.2 (b)].

The another form of emission is stimulated emission. [Refer Fig. 9.2 (c)]. When the atom is in excited state, before coming it to ground state, the atom is triggered due to an action of incident photon. The interaction between the excited atom and incident photon can trigger the excited atom to make a transition to ground state. The transition generates another photon which is identical to the incident photon (same phase i.e. coherent, same wavelength. i.e. monochromatic, same direction).

This process of forced emission of photons due to incident photon is called as stimulated emission.

9.3 ABSORPTION OR STIMULATED ABSORPTION

When a photon of energy $h\upsilon = E_2 - E_1$ is incident on an atom then the atom gets excited i.e. moves from lower energy level E_1 to higher energy level E_2.

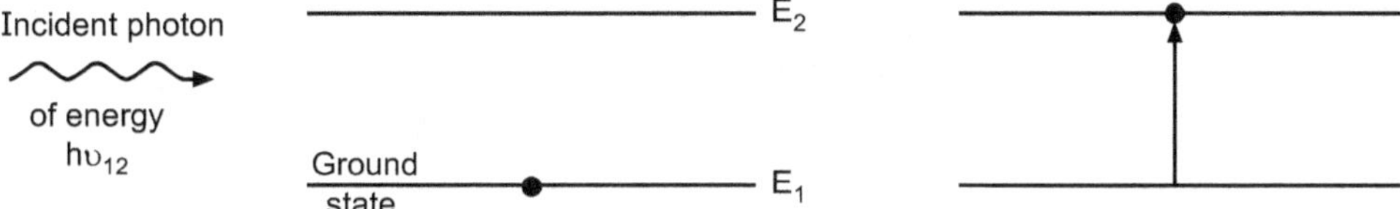

Fig. 9.3 : Excitation due to stimulated absorption

Here, the atom in the ground state absorbs the energy of incident photon and gets excited (stimulated) towards higher energy level E_2. This process is known as stimulated absorption.

9.4 SPONTANEOUS EMISSION

After completion of life time, the excited atom comes to ground (lower) energy state spontaneously (on its own accord) emitting a photon $h\upsilon$. This is known as spontaneous emission.

Fig. 9.4 : Spontaneous emission

The spontaneous emission depends on type of particle and type of transition but is independent of outside circumstances. Radiations which are emitted spontaneously are random in directions, random in phase. Thus, radiation in this case is a random mixture of quanta having different wavelengths, different phase. Thus, such a radiation is incoherent and has a broad spectrum.

9.5 STIMULATED EMISSION

When the atom is in excited state, before coming it to ground (lower) state the atom is triggered due to an action of incident photon. The interaction between the excited atom and incident photon can trigger the excited atom to make a transition to ground state. The transition generates another photon which is identical to the incident photon (same phase i.e. coherent, same wavelength i.e. monochromatic, same direction).

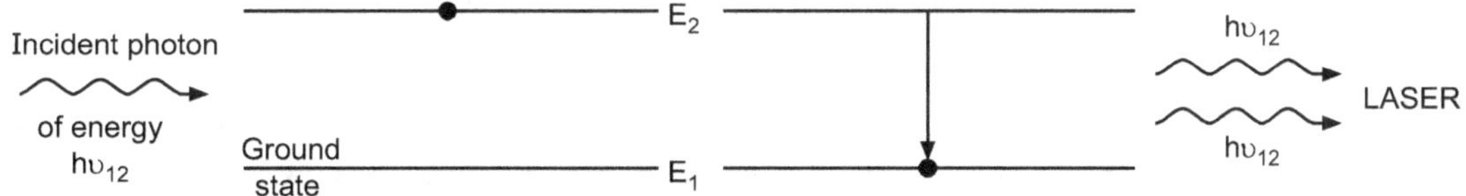

Fig. 9.5 : Stimulated emission

This process of forced emission of photons due to incident photon is called as stimulated emission.

Metastable State :

Usually the number of excited atoms in a system are less than that of non-excited atoms. The atoms prefer ground state for longer (unlimited) time. On the contrary, they remain in excited state for an limited time called life time. The life time of hydrogen is 10^{-8} sec. However, there exists such an excited state in which life time is greater than 10^{-8} sec. This state is called as metastable state. The relaxation time for metastable state is 10^{-6} to 10^{-3} seconds. This state plays an important role in laser.

Ordinary excited state :

Atom remains in excited state for very very small time (10^{-8} sec) and comes down to ground state immediately.

Metastable excited state :

Atom relaxes in excited state for long time (10^{-3} sec) and then comes down to ground state.

9.6 POPULATION INVERSION

Population means number of active atoms occupying an energy state.

Usually population of lower energy (ground) level is high and that of higher energy level is low as shown in Fig. 9.6.

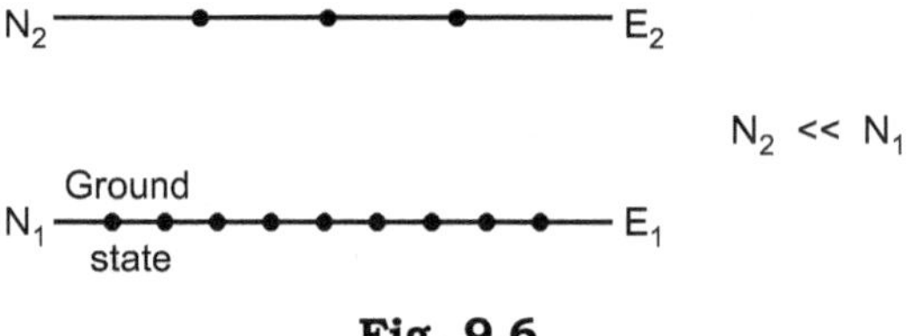

Fig. 9.6

In order to produce stimulated emission properly, population of higher excited state should be greater than that of low energy (ground) state.

Making population of higher energy level more than that of ground state is called population inversion or i.e. making $N_2 >> N_1$ is called as population inversion or inverted population as shown in Fig. 9.7.

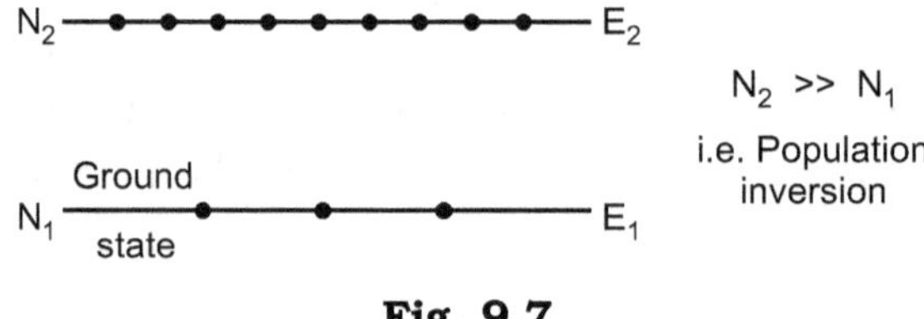

Fig. 9.7

Population inversion : Definition : Making population of active electrons in higher energy level (N_2) more than the population of active electrons in lower energy level (N_1) is called population inversion.

A system in which population inversion is achieved is called as active system.

The process of raising the atoms from lower energy state to higher excited energy state is called as pumping.

9.7 PUMPING METHODS

Pumping : Definition : The process of raising the atoms from lower energy state to higher energy excited state is called as pumping.

Optical pumping : Definition : The process of raising the atoms from lower energy state to higher energy excited state using light medium is called as optical pumping.

There are several methods to achieve population inversion, which is necessary for laser action to take place.

I. **Optical pumping** is generally used for solid-state lasers.

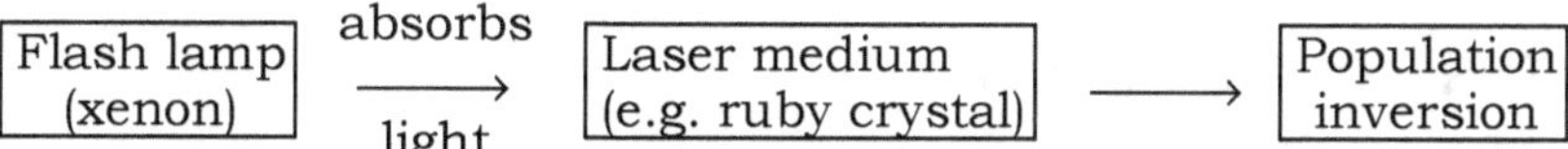

II. **Direct electron excitation (Electric pumping) :**

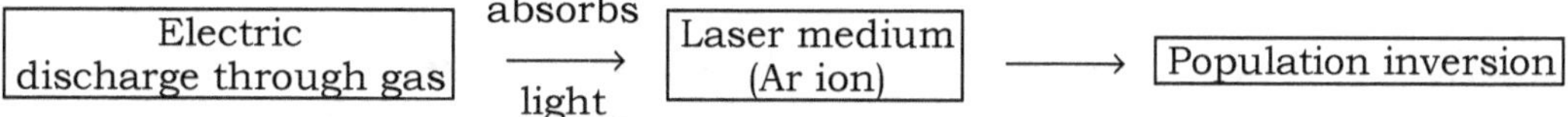

III. **Inelastic atom-atom collision :**

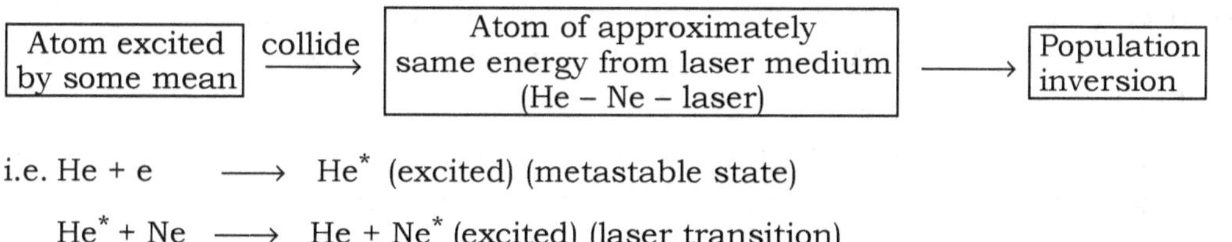

i.e. He + e $\longrightarrow$ He* (excited) (metastable state)

He* + Ne $\longrightarrow$ He + Ne* (excited) (laser transition)

IV. **Chemical reaction (chemical pumping)**

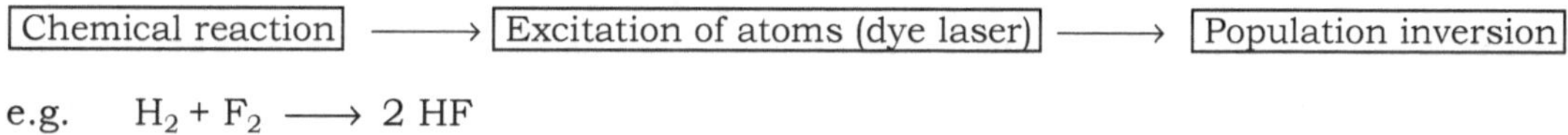

e.g. $H_2 + F_2 \longrightarrow 2\,HF$

This reaction gives hydrogen fluoride molecule in excited state.

9.8 OPTICAL PUMPING (THREE ENERGY LEVEL LASER SYSTEM)

Proper lasing action can be produced using three energy level laser system than that of two energy level laser system.

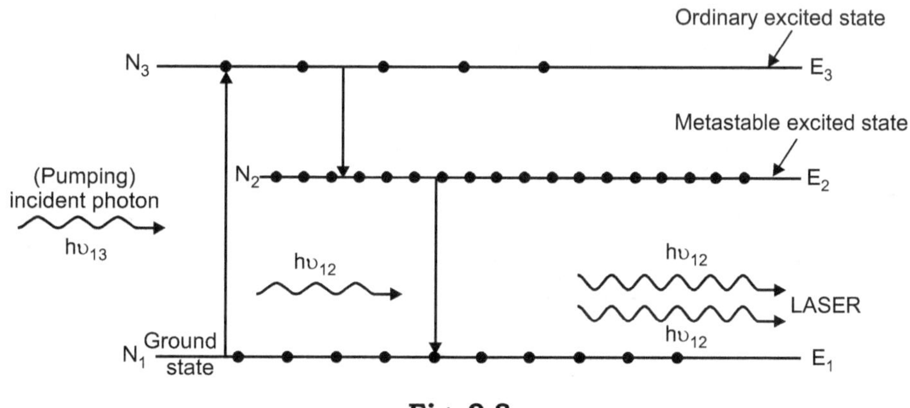

Fig. 9.8

Optical pumping i.e. photon of energy $h\upsilon_{13}$ is incident as shown because of which atoms get excited from energy level E_1 to energy level E_3. Here atoms relax for very very short time (i.e. life time is less than 10^{-9} sec) and hence atoms make transition to energy level E_2. This transition is not in the visible range. Energy level E_2 is a metastable (has life time 10^{-6} to 10^{-3} sec). Hence, atoms relax here for longer time. Hence, population of energy level E_2 becomes more than that of E_1 i.e. making $N_2 \gg N_1$ required population inversion is done.

If atom is triggered due to an action of an incident photon of energy $h\upsilon_{12}$, then excited atom gets stimulated and follows downward transition from E_2 to E_1. During this transition, the atom emits a photon along with incident photon in the same direction, which produces unidirectional, intense, monochromatic and coherent laser radiations.

9.9 PROPERTIES OR CHARACTERISTICS OF LASER

1. Coherence (same phase) : Laser light is perfectly coherent in nature. i.e. the waves are exactly in phase with one another. The emitted photon after getting triggered is exactly in phase with incident photon.

2. Monochromaticity (same wavelength or frequency) : The laser light is perfectly monochromatic. The light emitted by laser is much more monochromatic than that of any conventional monochromatic source.

3. Unidirectionality (sharp focus) : Conventional light source emits light in all directions. But laser emits (spreads) light in one direction. The width of laser beam is very narrow and can travel to long distance without spreading. Hence, it can be focussed sharply.

4. High intensity (extreme brightness) : Since emitted photon and incident photon are in same phase, laser light is much more brighter (intense) than that from any of the conventional source. It can vaporize hardest metal.

9.10 HELIUM-NEON LASER (HE-NE GAS LASER)

The gas atoms are characterized by sharp energy levels as compared to the solid. For continuous laser beam, gas lasers are used. The advantages of gas lasers are high monochromaticity and stability of frequency. He-Ne laser is commonly used to read bar codes.

Construction :

It consists of fused quartz tube having length of about 80 cm and diameter of about 1.5 cm. This tube is filled with a mixture of helium (He) and neon (Ne) gas. The mixture consists 10 parts of He and 1 part of Ne. Hence, there is majority of helium atoms (90%) and minority of neon atoms (10%). Perfect reflector is fixed at one end and partial reflector is fixed at other end of the tube. The active material is excited by means of high frequency generator.

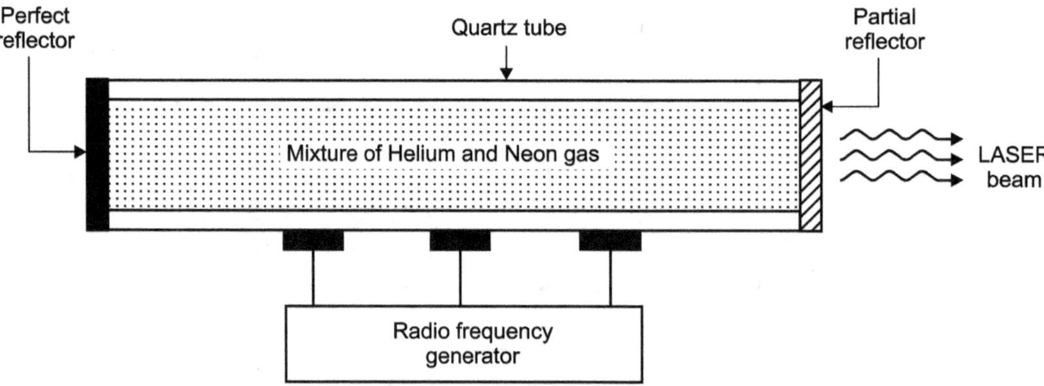

Fig. 9.9 : He-Ne gas laser

Working :

An electric discharge is produced in the gas by means of electrodes outside the tube connected to a source of high frequency alternating current. Collisions with electrons from the discharge excite He and Ne atoms to metastable states 20.61 eV and 20.66 eV respectively.

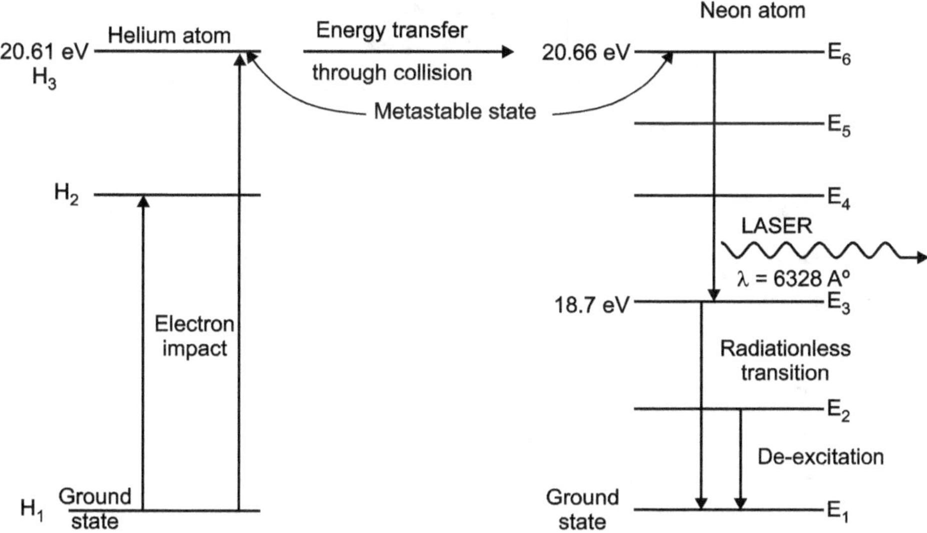

Fig. 9.10 : Energy level diagram for He-Ne laser

As the neon higher levels E_4 and E_6 are closed to excited energy levels H_2 and H_3 of helium, hence the probability of helium atoms transferring their energy to excited neon atoms through collision is greater. Thus, the purpose of helium atom is to help to achieve population inversion. Actual lasing atoms are neons.

The laser transition in Ne is from metastable state at 20.66 eV to an excited state at 18.7 eV. This is the laser whose narrow red beam is used to read bar codes.

9.11 APPLICATIONS OF LASER

- Lasers are used to read bar code in shopping malls, library etc.

- Lasers are used for **engraving** and **embossing** of printing plates e.g. number plate, name of company, monogram of the company.

- Lasers are used in cutting, drilling (peening) and welding metals.

- Lasers are used in surgery for destroying kidney stone, cancer cells.

- In military, lasers are used to direct the weapons towards the target.

- Lasers are used to measure distance of the planets. e.g. distance of moon from the earth.

- Lasers are used in holography, printers.

- Lasers are used by dentists for dental surgery.

- Lasers are used for 3D laser scanners – analyse the real world object – collected data can be used to construct digital 3-dimensional models.

- Laser pointer or laser pen – Laser pen is a small portable and visible laser designed to highlight some interesting part by focussing small bright spot of coloured light on to it.

- Lasers are used in eye surgery, soft tissue laser surgery, cosmetic surgery (e.g. removing tatoos, scars, stretch marks, wrinkles, sunspots etc.)

SUMMARY

- The term LASER stands for Light Amplification by Stimulated Emission of Radiation.

- Laser has remarkable properties like coherency, monochromaticity, unidirectionality and extreme brightness.

- When photon is incident on an atom, then atom absorbs energy and gets excited.

- The limited time for which an atom remains in excited state is called as life time.

- There are two types of emission :

  (1) Spontaneous emission.

  (2) Stimulated emission.

- The process of forced emission of photon due to incident photon is called as stimulated emission.

- Ordinary excited state : Atom remains in excited state for very small time and comes down to ground state.

- Metastable excited state : Atom relaxes in excited state for long time and then comes down to ground state.

- Making population of higher energy level more than that of ground state is called as population inversion.

- The process of raising the atoms from lower energy state to higher excited energy state is called as pumping.

EXERCISE

1. Give full form of LASER.

2. What do you mean by lasers ?

3. Define spontaneous emission and explain.

4. Define stimulated emission and explain.

5. Define population inversion.

6. Define optical pumping.

7. Define active system.

8. Explain working principle of laser.

9. Explain population inversion.

10. Explain optical pumping.

11. What is ordinary excited state and metastable state ?

12. State and explain properties of laser.

13. Explain construction, working of He-Ne gas laser with diagram.

14. Explain mechanism of production of laser.

15. Define the term life time of state during production of laser.

16. Draw energy level diagram for He-Ne laser.

17. State applications of laser.

10

CHAPTER

X-RAYS

10.1 Introduction

10.2 Production of X-rays using Coolidge Tube

10.3 Two Types of X-rays

10.4 Properties of X-rays

10.5 Applications of X-rays

Summary

Important Formulae

Solved Examples

Exercise

Problems for Practice

10.1 INTRODUCTION   (Nov. 18)

The X-rays were discovered by the scientist Rontgen in the year 1885. While he was performing experiments on discharge of electricity through gases in a discharge tube, he found that when fast moving electrons are stopped by a solid target, some unknown radiations were produced. These radiations were named by him as X-rays. They are electromagnetic radiations of very short wavelength (10^{-10} to 10^{-11} m). X-rays with longer wavelengths are called *soft X-rays* and those with small wavelengths are called *hard X-rays*.

10.2 PRODUCTION OF X-RAYS USING COOLIDGE TUBE

Principle :

When fast moving electrons (e^-) are suddenly stopped, then X-rays are produced.

The modern X-ray tube used for the production of X-rays is known as Coolidge tube. It consists of highly evacuated hard glass bulb with a cathode (filament) and anti-cathode or anode (as shown in Fig. 10.1).

(10.1)

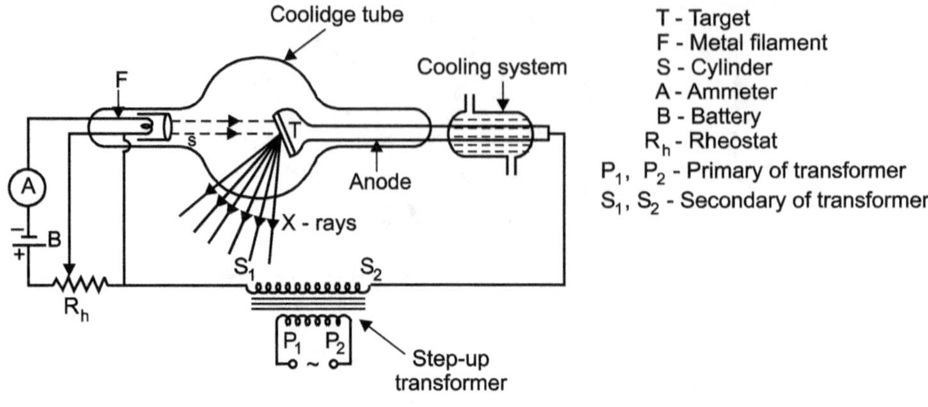

Fig. 10.1

The cathode i.e. metal filament is surrounded by molybdenum metal cylinder kept at negative potential to the filament. Hence, the electrons emitted from the filament are concentrated into fine beam of electrons.

The target T consists of copper block in which a piece of tungsten or molybdenum is fitted. The anode should possess following properties.

1. **High melting point :** So that it is not melted due to bombardment of fast moving electrons which produces large amount of heat.

2. **High atomic weight :** To produce the hard X-rays.

3. **High thermal conductivity :** To carry away the generated heat.

The target is placed at an angle of 45° with the path of electron beam.

When the cathode (filament) is heated by an electric current, it produces electrons due to process known as *thermionic emissions*. This beam of electrons is then focused on the anode (target). The electrons from the cathode are accelerated by application of high voltage between the cathode and anode by using step-up transformer. When these fast moving electrons are suddenly stopped by the tungsten anode, they loose their kinetic energy and X-rays are produced from the target. Some amount of K.E. is converted into large amount of heat.

By controlling the filament current, the thermionic emission of electrons and hence the intensity of X-rays can be controlled. The penetrating power of X-rays determines the quality of X-rays which can be controlled by changing voltage between cathode and anode.

The X-rays of high penetrating power and higher frequency are called hard X-rays and those with low penetrating power and low frequency are called soft X-rays.

Intensity of X-ray depends on filament current. Penetrating power of X-ray depends upon potential difference (P.D.) between cathode and anode.

10.3 TWO TYPES OF X-RAYS

10.3.1 Characteristic X-Rays (Line Spectrum)

We know that electrons in the outer orbit are loosely held by nucleus.

Inner orbit electrons being closer to nucleus are more tightly held by nucleus. In this case, incident electron penetrates deep into an atom. They knock out one of the inner shell electron of target atom, which creates a vacancy i.e. hole (lack of electron) in that orbit say K shell. Later on electron from higher orbit say L shell jumps into the hole of K shell. Then the energy difference $E_K - E_L$ is emitted in the form of characteristic X rays.

The K_α spectral line originates when an electron from L shell jumps into the hole of K shell.

The K_β spectral line originates when an electron from M shell jumps into the hole of K shell.

When electron from higher orbit i.e. L, M, N jumps into K shell which gives rise to K series. i.e. K_α, K_β, … .

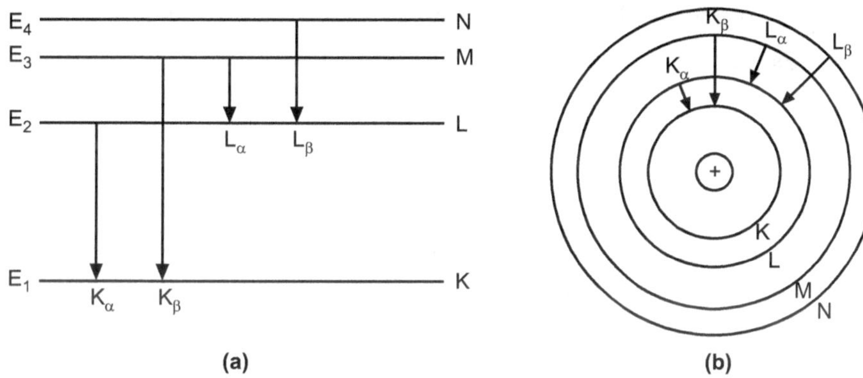

Fig. 10.2

When electron from higher orbit jumps into the hole of L shell, it gives rise to L series i.e. L_α, L_β …….

10.3.2 The Continuous X-Ray (Spectrum)

It is produced due to the retardation of fast moving electrons when they are deflected while passing near the nucleus of an atom of the target material.

Consider an electron of initial kinetic energy KE_0 which collides with one of the target atom, after collision, its kinetic energy reduces (loses) and which comes out (converted) in the form of X-ray photon.

Then this electron having less kinetic energy gets deflected and will have second collision with target atom, generating a second X-ray photon. The energy of second X-ray photon is less than that of first photon.

Thus, electron speed goes on decreasing. It may have third collision with target atom, generating third X-ray photon of further less energy.

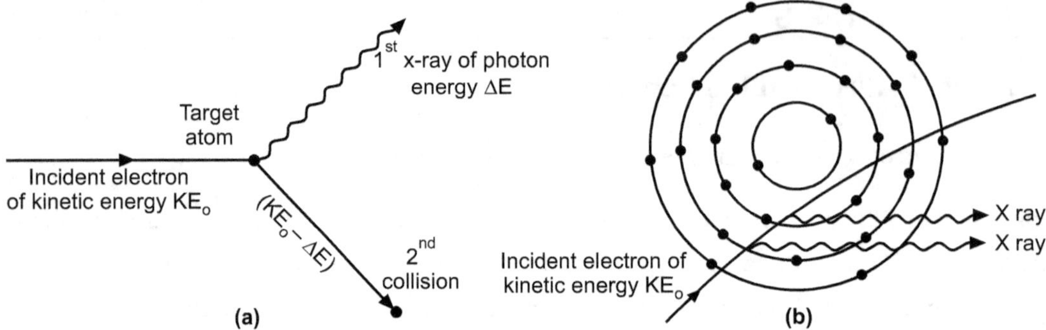

Fig. 10.3

This electron scattering process can continue until the electron has approximately zero velocity.

Emitted energy varies continuously since the loss in kinetic energy of electron is continuous, hence X-rays emitted are called as continuous X-rays or white X-rays.

Minimum wavelength :

A feature of this spectrum is the sharply defined cut-off wavelength λ_{min}, below which the continuous spectrum does not exist.

Cut-off wavelength, $\lambda_{min} = \dfrac{hc}{KE_o}$

or $\lambda_{min} = \dfrac{hc}{eV}$

where, h = Planck's constant

c = Speed of light = $3 \times 10^8 \, m/s$

KE_o = Kinetic energy of incident photon

e = Charge on electron

= 1.6×10^{-19} coulomb

V = Applied voltage

Maximum frequency :

We have, Velocity of wave = Frequency × Wavelength

∴ $f_{max} = \dfrac{Velocity}{Wavelength}$

$f_{max} = \dfrac{c}{\lambda_{min}}$

where, c = Speed of light

λ_{min} = Cut-off wavelength

10.4 PROPERTIES OF X-RAYS   (Nov. 18)

1.   They are electromagnetic waves of very short wavelength.

2.   They travel with speed of light ($c = 3 \times 10^8$ m/s).

3.   X-rays affect photographic plates.

4.   They produce fluorescence in many substances e.g. zinc sulphide.

5.   They can be reflected or refracted under certain conditions.

6.   They are not deflected by magnetic or electric field.

7.   They have high penetrating power and hence can pass through many solids.

8.   They produce small amount of ionization in the gases through which they pass.

9.   They produce photoelectric effect.

10. X-rays kill some form of animal cells.

10.5 APPLICATIONS OF X-RAYS

(A) Engineering Applications :

1.   X-rays are used to detect the cracks in the body of an aeroplane or motor car.

2.   X-rays are used to detect manufacturing defects in rubber tyres or tennis balls in quality control.

3.   X-rays are used to detect flaws or cracks in metal jobs.

4.   They are used to distinguish real diamond from duplicate one.

5.   X-rays are used to detect smuggling gold at air port and dock (ship) yard.

6.   They are used to detect cracks in the wall (civil engineering).

7.   X-ray radiography is used to check the quality of welded joints.

(B) Medical Applications :

8.   X-rays are used in surgery to detect bone fractures. The reason is penetration power of X-rays through muscle (skin) and bone is different.

9.   X-rays are used to cure skin diseases and destroy tumours.

10. X-rays are used to cure diseases like cancer.

11. X-rays are used to detect bullet's position inside the body.

(C) Scientific Applications :

12. X-rays are used to study structure of crystal (e.g. whether hexagonal, rhombus, ...). The structure of an alloy is studied with the help of diffraction of X-rays to determine the crystal form.

13. They are used in chemical analysis and for determination of atomic number of chemical elements.

14. They are used to study structure of substances like cellulose, rubber, plastic.

SUMMARY

- When fast moving electrons are suddenly stopped, then X-rays are produced.

- Characteristic X-rays and continuous X-rays are the two types of X-rays.

- X-rays have high penetrating power, this is the remarkable feature.

IMPORTANT FORMULAE

$$\lambda_{min} = \frac{hc}{eV}$$

where, h = Planck's constant

$$= 6.62 \times 10^{-34} \, Js$$

c = Velocity of light = $3 \times 10^8 \, m/s$

e = Charge on electron = $1.6 \times 10^{-19} \, C$

V = Applied voltage

or $$\lambda_{min} = \frac{12400 \times 10^{-10}}{V} \, m$$

$$= \frac{12400}{V} \, A^\circ$$

$$E = h\upsilon$$

$$1 \, eV = 1.6 \times 10^{-19} \, J$$

where　$1 \, A^\circ = 1 \times 10^{-10} \, m$

E = Energy of photon or spectrum

υ = Frequency of photon

SOLVED EXAMPLES

Example 10.1 : *An X-ray tube works on 50 kV. What will be the wavelength of X-rays emitted in it ?*

Solution : Given : $V = 50 \, kV = 50 \times 10^3 \, V$,　$\lambda = ?$

We have,　$$\lambda = \frac{12400}{V} \, A^\circ = \frac{12400}{50 \times 10^3}$$

$$\boxed{\lambda = 0.248 \, A^\circ}$$

Example 10.2 : *Find minimum wavelength and maximum frequency of X-ray produced by an X-ray tube working on 50 kV.*

($h = 6.62 \times 10^{-34} \, Js$, Velocity of light = $3 \times 10^8 \, m/s$, $e = 1.6 \times 10^{-19}$ coulomb)

Solution :　$$\lambda_{min} = \frac{h\,c}{eV} = \frac{(6.62 \times 10^{-34})\,(3 \times 10^8)}{(1.6 \times 10^{-19})\,(50 \times 10^3)}$$

$$\boxed{\lambda_{min} = 0.248 \, A^\circ = 0.248 \times 10^{-10} \, m}$$

$$f_{max} = \frac{c}{\lambda_{min}} = \frac{3 \times 10^8}{0.248 \times 10^{-10}}$$

$$\boxed{f_{max} = 120 \times 10^{17} \, Hz}$$

Example 10.3 : *An X-ray tube operated at 30 kV emits a continuous X-ray spectrum with short wavelength limit, λ_{min} = 0.414 A°.*

Calculate the charge on electron if h = 6.63×10^{-34} Js and c = 3×10^8 m/s

Solution : Given : λ_{min} = 0.414 A° = 0.414×10^{-10} m

$$V = 30 \text{ kV} = 30 \times 10^3 \text{ V}$$

$$e = ?$$

We have, $\qquad \lambda_{min} = \dfrac{hc}{eV}$

$\therefore \qquad e = \dfrac{hc}{\lambda_{min} V} = \dfrac{(6.63 \times 10^{-34}) \times (3 \times 10^8)}{(0.414 \times 10^{-10}) \times (30 \times 10^3)}$

$$\boxed{e = 1.6 \times 10^{-19} \text{ joules}} \text{ or } \boxed{e = 1 \text{ eV}}$$

Example 10.4 : *Calculate the minimum applied potential required to produce X-rays of 0.51 A° wavelength.*

Solution : Given : λ = 0.51 A° = 0.51×10^{-10} m, V = ?

We have, $\qquad \lambda_{min} = \dfrac{hc}{eV}$

$$\lambda = \dfrac{12400 \times 10^{-10}}{V}$$

$\therefore \qquad V = \dfrac{12400 \times 10^{-10}}{\lambda} = \dfrac{12400 \times 10^{-10}}{0.51 \times 10^{-10}}$

$\therefore \qquad \boxed{V = 24313 \text{ volt}}$

Example 10.5 : *Find the shortest wavelength λ_{min} that can be emitted by the sudden stoppage of fast moving electrons when they strike the screen on T.V. tube operating at 40 kV.*

Solution : Given : V = 40 kV = 40×10^3 V

$$\lambda_{min} = ?$$

$$\lambda_{min} = \dfrac{hc}{eV} = \dfrac{12400}{V} \text{ A}° = \dfrac{12400}{(40 \times 10^3)}$$

$\therefore \qquad \boxed{\lambda_{min} = 0.31 \text{ A}°}$

Example 10.6 : *The energy of X-ray spectrum is 3.3 eV. Find its frequency.*

(Given : h = 6.6×10^{-34} Js and 1 eV = 1.6×10^{-19} joules)

Solution : Given : E = 3.3 eV

$$= 3.3 \times 1.6 \times 10^{-19} \text{ joules}$$

We have, $\qquad E = h\upsilon$

$\therefore \qquad \upsilon = \dfrac{E}{h} = \dfrac{3.3 \times 1.6 \times 10^{-19}}{6.6 \times 10^{-34}}$

$$\boxed{\upsilon = 8 \times 10^{14} \text{ Hz}}$$

EXERCISE

1. Describe how X-rays are produced by using modern Coolidge X-ray tube.

2. Draw neat labelled diagram of modern X-ray tube.

3. Give four applications of X-rays.

4. State four important properties of X-rays.

5. Explain construction and working of Coolidge tube.

6. State uses of X-rays.

PROBLEMS FOR PRACTICE

1. Find the minimum wavelength of X-rays produced by an X-ray tube operated at 1000 kV. (Given : h = 6.63×10^{-34} Js, e = 1.6×10^{-19} C and c = 3×10^8 m/s)   (**Ans.** 0.01237°)

2. An X-ray tube is operated at 18 kV. Calculate the minimum velocity of electron bombarding at anticathode.   (**Ans.** 8×10^7 m/s)

3. An X-ray is operated at 20 kV. Calculate the minimum wavelength of X-rays emitting from it.

(**Ans.** 0.62 A°)

□□□

INTRODUCTION TO NANOTECHNOLOGY

11.1   Introduction
11.2   What is Nano, Nanoscale, Nanometer, Nanoparticle, Nanostructured material
11.3   Applications of Nanotechnology
        Exercise

11.1 INTRODUCTION

AMAZING - INTERESTING - NOVELTY (IDEA ABOUT NANOTECHNOLOGY)

- We know that taste of powdered sugar is different than taste of sugar cube. i.e. size matters.

- If we take a cube of gold having each side 1 cm (i.e. 1 cm³) and note down its properties e.g. colour - yellow shiny. Now, cut this cube into half slice along length, breadth and height.

- Now, we will have eight cubes of dimensions 0.5 cm each side.

- Now, take this small cube and again cut into eight equal cubes. If we go on doing this process, **then below certain size** (boundary), the colour of the tiny gold particles goes on changing. It may be orange, purple, red or greenish depending on their size. i.e. NOVEL (AMAZING) properties. This will happen when we reach the **nanoscale**. Everything of gold will change i.e. colour, melting point, chemical properties, strength, resistance etc. e.g. a nano wire may not obey Ohm's law.

- This (BOUNDARY) is not fixed for every material. It changes from material to material. The last few steps of this cutting process come under kind of **nanofabrication** i.e. **nanoscale manufacturing**.

- We go from the cube towards nano-particles. This kind of **nanofabrication** is called the **top-down nanofabrication**.

- But if we start from atoms and go upto nanoscale particles then it is called **bottom-up nanofabrication**. These tiny gold particles may be called as **quantum dots** or **nanodots** because they are nearly dot shaped in the nanoscale.

11.2   WHAT IS NANO, NANOSCALE, NANOMETER, NANOPARTICLE, NANOSTRUCTURED MATERIAL ? (Nov. 18)

1. **Nano :**   From the Greek word nanos - meaning is dwarf i.e. short man.

    Nano is the prefix used in the metric system to mean 10^{-9} or $\frac{1}{10^9}$ or one billionth $\left(\text{i.e. } \frac{1}{1,000,000,000}\right).$

2. **Nanometer is 10^{-9} meter or one billionth of meter i.e. 1 nm = 1×10^{-9} m.**

3. **Nanoscale : The scale which enables us to measure dimensions in nanometer i.e. 10^{-9} m range.**

   Some Examples :

   (i)   Dust from desert ~ 100 nm.

   (ii)   Human hair width ~ 60,000 nm.

   (iii)   Virus - 10 to 60 nm.

   (iv)   Paint pigments - 80 nm to 100 nm.

   (v)   Nanotube - 1 nm to 5 nm $\times$ 10 mm.

   (vi)   Quantum dots - 5 nm to 20 nm.

   (vii) 10 hydrogen atoms in line = 1 nm.

4. **Nanoparticles :** There is no fixed definition for a nanoparticle because the boundary below which properties of material change, is not fixed for every material. i.e. the size at which materials display different properties as compared to the bulk material is material dependent. Normally, it is below 100 nm.

   Nanoparticle is a particle having dimensions of the order of less than 100 nm.

   Novel properties that differentiate nanoparticles from the bulk material happens below the critical length scale of under 100 nm.

   Properties of nanoparticles are different than the properties of bulk material (same material). Thus the size of particles is an important feature of nanoparticle.

5. **Nanostructured material (NSM) : Nanomaterial is only the building block of the nanosystem. Nanostructured materials are those with at least one dimension falling in the nanometer scale and include nanoparticles (including quantum dots)** or nanostructured materials are the materials with microstructure characteristic length, the scale of which is of the order of 1-100 nm.

   Examples :

   Conventional nanomaterials : Nanofibers, inorganic nano-particles, nano clays, multiwalled nanotubes etc.

6. **Molecular nanomaterials :** Thin film organized nanostructure, designer protein, nanostructured polymers, biomimetic materials, organic nanoparticles, single walled nanotubes, nano rods, nano wires, carbon nanotubes, nanocomposite structures.

11.3 APPLICATIONS OF NANOTECHNOLOGY   (Nov. 18)

- Nanotechnology is the science of engineering matter at the atomic and molecular scale. It is the manipulation, precision placement, measurement, modeling or manufacture of less than 100 nm scale matter. It is the study of the control of matter on an atomic and molecular scale. It deals with the structure of size 100 nm or smaller in at least one dimension.

- Nanotechnology or molecular manufacturing consists of manipulating matter on atom by atom or molecule by molecule basis to attain the designed configurations.

Nanotechnology (Definition) : The design, characterisation, production and application structures, devices and systems by controlled manipulation at nanometer scale (atomic, molecular) that produces structures with at least one novel/superior property.

11.3.1 Applications of Nanotechnology in Electronics (Nanoelectronics)

Nanoelectronics increases the capabilities of electronic devices and reduces their weight and electric power consumption.

- **Improvement in display screens on electronic devices** e.g. computer monitor, laptop screen, mobile screen display, which involves reducing power consumption while decreasing weight and thickness of the screen.

- Flat computer monitors or TV screens or coating used on screens may be of nanoparticles which improve performance.

- **Increasing the density of memory chips :** i.e. volume of material of memory chip is reduced to a great extent and the capability of memory increased i.e. amount of information stored.

- Nanoradios have been developed.

- Use of nanomaterial such as nanowire in place of traditional CMOS components.

- Reducing the size of transistors used in integrated circuits ICs, because of which some kind of mobiles which are as good as computer are possible in future.

- **Some nanotechnology electronic devices are**

    (i) **Single Electron Transistor (SET)** - which involves transistor operation based on a single electron (e^-).

    (ii) **Single Molecule Transistor :** These schemes make use of molecular self assembly designing the device components to construct a large structure or complete system.

    (iii) **Spin valves** used in computers to read disks.

    (iv) **Magnetic Tunnel Junction (MTJ) :** These devices need less power, they are cheaper, compact in size and faster in speed (access).

- **Reducing the size of transistors used in integrated circuits ICs :** Besides being small and allowing more transistors to be placed on a single chip, the uniform and symmetrical structure of nanotubes allows a higher electron mobility (faster electron movement in the material), a higher dielectric constant (faster frequency) and symmetrical electron/hole characteristics.

11.3.2 Applications in Automobiles

- **Nanoparticle paints :** Nanoparticle paint applies a layer of colour which is smooth and shiny. A car painted by this technique is very easy to wipe (clean) and reshine.

- One of the problems of using hydrogen as a fuel is its storage problem. Hydrogen has to be stored in a cylinder at high pressure. If a metal cylinder is used, then the weight of the vehicle increases and hence there is reduction in milleage (km/litre). This problem may be solved by storing this hydrogen in nanocylinders of carbon nanotubes.

- **Nanotechnology to reduce air pollution :** A catalyst is used to accelerate a chemical reaction effectively at low temperature. Catalysts made from nanoparticles have **greater surface area** to interact and hence it is efficient catalyst which produces **less harmful gases**.

- Tyres of car made using nanotechnology (nanoprocess) will be of light weight which may increase milleage (km/litre) of a car.

- Small motors required for a CD player, wiper movement made from nanotechnology will reduce the size and weight and hence reduce the power consumption.

- While manufacturing glass, nanoparticles are used so that glass cleaning becomes easier.

11.3.3 Applications in Medical (Nanotechnology in Medicine)

- Applications of nanotechnology are being developed in advanced drug delivery system. Nanoparticles are used to deliver drugs, heat and light to specific cells in the human body. Target and controlled delivery of drugs is possible which allows treatment of injury or disease within the targeted cell, thereby minimizing the damage to healthy cells in the body.

- **Protein and Peptide delivery :** Protein and peptide exert multiple biological action in human body and are used for treatment of various diseases and disorders. Nanoparticles are used to deliver protein and peptide.

- **Treatment of Cancer :** Nanoparticles or cancer drugs are used to treat cancer by destroying cancer cells and not healthy cell.

- **Used in surgery to weld (join) two pieces together :** During operation a certain part is cut to operate the inner organs properly. These two pieces are placed together and greenish liquid containing gold-coated nanoshells is dribbled along the seam. An infrared LASER is traced along the seam causing the two sides to be welded (joined) together. This could solve difficulties and blood leaks when the surgeon tries to restitch the arteries cut during a kidney or heart transplant.

- The future of nanotechnology in medicine is referred to as nanomedicine which involves nano-robots (e.g. computational genes) introduced into the body to repair or detect damages, infections and cell repairy.

- **Molecular machine :** Using drugs and surgery, doctors can only encourage tissues to repair themselves. With molecular machine, there will be more direct repair. Access to these cells is possible because a doctor can stick the needle into the cell without killing them. Thus, molecular machines are capable of entering into the tissue.

11.3.4 Nanotechnology in Textile

Using nanotechnology, special threads are produced which are durable (long life), shiny. Clothes produced using nanotechnology are shiny, give pleasant look and also give the feeling of cotton. Silver nanoparticles can be used in washing machine to make clothes germ free, in this silver coated plate (chip) is used. Its life may be for approximate 5000 washings.

11.3.5 Application in Cosmetics

Nano-based colour is used in hair creams or gels which gives a shiny effect. **Zinc oxide nanoparticles absorb ultraviolet light and protect the skin from sunlight** and hence is used in face cream or powder. Because of their nanosize these particles easily fill up the microgaps on the face and give a smooth effect. Scattering of light from these nanoparticles of the cream gives wrinkle-free look.

11.3.6 Environmental Applications

- **Catalyst** can be used to enable chemical reactions at low temperature more efficiently. Catalysts made using nanoparticles have greater surface area to interact with the chemicals. Because of larger surface area, more chemicals interact with the catalyst which makes the catalyst more effective.

- **Nanostructured membranes** are under development to separate carbon dioxide from industrial plant exhaust flow.

Nanotechnology applications under development :

- Removal of organic compounds from industrial smoke stacks.

- Reducing the amount of platinum used in catalytic converter.

- Removal of carbon dioxide from smoke stacks using nanostructured membranes and carbon nanotube based membranes.

- Nanoparticle-based **sensors** are used to detect toxic ions, poisonous gases, pesticides and hence to control pollution.

- Nanoparticle-based sensors are used in water purification system.

11.3.7 Nanotechnology in Space and Defence

- Nanotechnology makes space flight more practical. Nanomaterials make light weight cable for space elevator and light weight solar cells which reduce amount of rocket fuel required and hence the cost.

- New materials combined with nanorobots can improve the performance of space suits and space ship. Light weight jackets are made using aerogels which are poor conductors of heat. Aerogel is a special form of solid of nanoporous material of very low density which reduces weight and gives comfort.

- Use of nanoparticles in polymer composites makes good radiation protectors.

- Researchers are working in following areas :

  (i) Use of carbon nanotubes to make the cable required for space elevator, reduces the weight and hence cost. Carbon nanotubes are molecular-scale tubes of graphite carbon which are **stiffest and strongest** fibres.

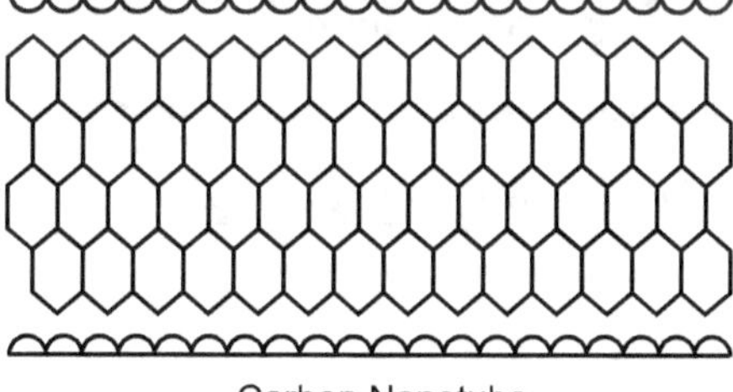

Carbon Nanotube

Fig. 11.1

  (ii) Use of materials made from carbon nanotubes reduces the weight of spaceship and increases the structural strength.

SUMMARY

- **Nano** (Greek word, meaning dwarf) is a prefix used in the metric system to mean 10^{-9} or $\frac{1}{10^9}$ or one billionth $\left(\text{i.e. } \frac{1}{1,000,000,000}\right)$.

- The scale which enables us to measure dimensions in nanometer i.e. 10^{-9} m range is called **nanoscale**.

- **Nanoparticle** is a particle having dimensions of the order of less than 100 nm.

- **Nanostructured materials** (NSM) are those with at least one dimension falling in the nanometer scale and include nanoparticles (including quantum dots).

OR

  These are the materials with microstructure characteristic length, the scale of which is of the order of 1-100 nm.

- **Nanotechnology :** The design, characterisation, production and application structures, devices and systems by controlled manipulation at nanometer scale (atomic, molecular) that produces structures with at least one novel/superior property.

- Nanotechnology is very useful in various fields like Electronics, Automobiles, Medical, Textile, Cosmetics, and also in Space and Defence.

EXERCISE

1. Define nanometer.

2. What is nanometer ?

3. What is nanoparticle ?

4. Define nanostructured material.

5. What is nanostructured material and list four examples of nanostructured materials ?

6. List the names of the fields where nanotechnology can be used.

7. State applications of nanotechnology in electronics.

8. State applications of nanotechnology in automobiles.

9. State applications of nanotechnology in medical field.

10. State applications of nanotechnology in textile industry.

11. State applications of nanotechnology in cosmetics.

12. How nanotechnology can be helpful in environmental field ?

13. State applications of nanotechnology in space and defence.

12

CHAPTER

NON-CONVENTIONAL SOURCES OF ENERGY

12.1 Introduction

12.2 Energy Resources

12.3 Non-Renewable Sources of Energy (Conventional Energy Sources)

12.4 Renewable Sources of Energy (Non-conventional Energy Sources)

12.5 Solar Energy

12.6 Wind Energy

12.7 Tidal Energy

12.8 Geothermal Energy

12.9 Biomass Energy

Summary

Exercise

12.1 INTRODUCTION

- *Energy* is crucial part in our life. Because any physical activity in this world, whether carried out by human beings or by nature, is caused due to flow of energy in one form or the other.

- Energy is required to do any kind of work.

- The word 'energy' itself is derived from the greek word 'en-ergon', which means 'in-work' or 'work content'. The work output depends on the energy input.

- The capability to do work depends on the amount of energy utilized.

- Energy is the most basic infrastructure input for economic growth and development of country.

- The proper use of energy sources requires consideration for technology's social impact with its economical growth as well as quality improvements.

12.2 ENERGY SOURCES

- Energy sources are forms of several energy. Energy sources can be classified follows :

1. Based on Use of Energy:

(a) Primary sources: These sources are first raw form of energy which cannot be used directly. These forms of energy are extracted, processed and converted into useful form as required by user.

Example: Coal, crude oil, sunlight, wind, running rivers etc.

(b) Intermediate sources: These are obtained from primary energy by one or more steps of conversion and are used as vehicle of energy.

Example: Electricity and hydrogen.

2. Based on Traditional Use:

(a) Conventional: The energy sources which are being traditionally used for many decade are called conventional energy sources.

Example: Fossil fuel, nuclear energy and hydro energy sources.

(c) Non-conventional: Energy sources which are available in large scale as compared to conventional energy sources are called non-conventional energy sources.

Example: Solar, wind, biomass etc.

3. Based on Availability:

(a) Non-renewable sources: Energy sources having limited availability and which cannot be used again after their consumption, are called non-renewable sources. These are also called as *finite sources*.

Example: Fossil fuels, uranium.

(b) Renewable sources: Energy sources which are renewed by nature again and again and the stock of energy is not affected by the rate of their consumption are called renewable sources.

Example: Solar, wind, biomass, ocean, geothermal, hydro etc.

4. Based on Commercial Application:

(a) Commercial energy source: The secondary usable energy forms such as electricity, petrol, diesel, gas etc. essential for commercial activities are called as commercial energy sources.

(b) Non-commercial energy sources: The energy which comes from nature and used directly without any conversion process is called a non-commercial source.

Example: Wood, animal dung, crop residue etc.

(c) Secondary sources: The form of energy which is finally supplied to a consumer for utilization is known as secondary energy.

Example: Electrical energy, thermal energy, chemical energy etc.

5. Based on Origin:

Energy resources are mainly identified and well known to every one.

(a) Fossil fuel energy

(b) Nuclear energy

(c) Hydro energy

(d) Solar energy

(e) Wind energy

(f) Biomass energy

(g) Geothermal energy

(h) Tidal energy

(i) Ocean thermal energy

12.3 NON-RENEWABLE SOURCES OF ENERGY (CONVENTIONAL ENERGY SOURCES)(Nov. 18)

- Non-renewable energy is, as its name suggests, energy that comes from a source that cannot be renewed. Or, in some case, energy that comes from a source that cannot easily be renewed.

- When we use non-renewable energy, we deplete the energy source and we cannot re-use the energy that we have used. Once it's gone, it's gone for good.

- Non-renewable energy sources are those sources that drain fossil reserves deposited over centuries. This results in depletion of these energy reserves. There are many countries, which have recorded significant reduction of these sources and are currently suffering from the side effects of drilling these energy reserves from deep underground. Examples of these countries include China and India.

- There are many places in the world that are experiencing fast degradation of non-renewable sources in terms of fossil fuels. Soon there will be none left if appropriate measures are not taken into consideration. This is a trend that has to be reversed if the world is to survive the degradation process that is going or happening at a much rapid pace. The main non-renewable energy sources are:-

- **Coal :** Coal is also composed of organic matter --- matter that decomposed in peat bogs, which then formed into carbon rock under immense pressure. Coal is generally highly combustible and the world's most-used source for electrical generation. However, burning coal releases massive amounts of carbon dioxide into the atmosphere, which is the primary factor in the greenhouse effect. In addition to being greenhouse gas source, coal cannot be reproduced.

- **Natural Gas :** Also a byproduct of decomposition, natural gas is mostly methane created as organic matter decays. Once extracted, natural gas is processed to remove everything but the methane. This produces a variety of other natural gases, such as ethane, propane and butane, which are also used as fuels. These natural gases are used to heat homes and businesses and fuel stoves. While it is a fairly clean-burning fossil fuel, it is a non-renewable energy source because it is a byproduct of thousands of years of decomposition within the earth's crust.

- **Petroleum :** Petroleum is extracted and turned into a variety of fuel sources including petrol or gasoline, diesel, propane, jet fuel, heating oil and paraffin wax. Also known as crude oil, this fuel source is non-renewable. Petroleum is made when organic matter settles in water that has lost its dissolved oxygen and is then compressed under immense heat and pressure for millions of years. There is no way for humans to reproduce this process for mass production either in nature or in a lab, so once mankind has used the current supply of petroleum, more will not be available for many centuries.

- **Nuclear Power :** Nuclear energy may get mentioned in the same breath with renewable power sources like wind and solar because it is clean-burning and therefore more environmentally sound than oil or coal. But nuclear energy is, in fact, a non-renewable source. The problem lies in the element that enables nuclear power: uranium. The element uranium is abundant, but only a certain type of uranium, U-235, is used to fuel nuclear power. U-235 must be extracted from mined and processed uranium. The processing produces only small amounts of U-235, making it rare and expensive. Man cannot reproduce this element; we have a limited natural supply.

12.3.1 Future of Non-Renewable Energy Sources

- Non-renewable energy sources have fueled the world's industrial complex for far too long. It has reached a point where the world is facing rapid starvation in this sector. There are also other associated effects too. With increased exploitation of these fossil fuels, there are many associated environmental effects like land pollution and air pollution, which in turn affect both animal and plant life. The far-reaching consequences of non-renewable sources are inexplicable and the trend has to be reversed soon before it is too late to do anything.

- Carbon is a major source of fuel in non-renewable energy sources. When combustion takes place, carbon is mixed with oxygen and forms carbon dioxide. It pollutes the environment and is responsible for global warming. In last few years, the concentration of carbon dioxide has only increased in the atmosphere. Not to mention, climate change, acid rain and change in seasons are some other effects that has been observed by many people. With so many problems, scarce resources and rising prices, these resources cannot be used for lifetime. The need of the hour is to look for some alternative sources of energy and protect our environment from such harmful gases.

12.3.2 Advantages of Non-renewable Energy

1. Non-renewable sources of energy such as coal and oil tend to provide us with more energy per unit than do renewable energy sources such as solar or wind energy.

2. Because they tend to come in solid, physical form such as lumps of coal or barrels of oil, non-renewable energy sources can usually easily be taken from place to place.

3. There are many profits to be made in the mining of coal, the selling of oil or the construction of natural gas pipelines.

4. Non-renewable energy sources are easy to burn, whether in a home or in a large factory.

5. Many non-renewable sources of energy (such as natural gas) are actually cheap per unit.

6. Extracting, transporting and refining non-renewable sources of energy – as well as creating the machinery for using them – creates plenty of jobs and thus can be one way of sustaining a country's economy.

12.3.3 Disadvantages of Non-renewable Energy

1. Mining coal, searching for oil, and building drills and pipes to extract and transport natural gas, are all very time consuming processes. This energy takes a lot of efforts to get hold of!

2. Burning coal, oil and natural gas releases large amounts of carbon dioxide and carbon monoxide which are major contributors to global warming. These chemicals destroy the ozone layer, make the oceans acidic and saturated with carbon, and make the air more difficult for animals to breathe and plants to flourish in.

3. Burning fossil fuels releases oxides (including sulphur oxide) which cause rain to become acidic. This is very harmful to wildlife and also erodes buildings.

4. Non-renewable energy sources like fossil fuels can emit carbon monoxide. This is dangerous to humans and can cause respiratory problems and death if inhaled.

5. These resources are not viable for future generations. That means we will not be able to base our lives on them forever. In fact, our reserves of many non-renewables may run out by the end of this century. Using non-renewable energy sources without taking steps to make our infrastructure, homes and factories ready to use renewable sources of energy could be said to be very selfish. Moreover, using up all the non-renewable sources of energy now, without leaving any for future generations, can also be said to be a selfish act.

6. Many non-renewable sources of energy are quite dirty, leaving soot and dirt on furnishings in the home. When used in factories, they release soot and other dirty substances into the air which can coat buildings and pavements, and make cities feel dirty and grimy.

12.4 RENEWABLE SOURCES OF ENERGY (NON-CONVENTIONAL ENERGY SOURCES)(Nov. 18)

12.4.1 Necessity of Renewable Energy

- The energy consumption of the world has been increasing with industrial growth.

- The fossil fuel is going to be exhaused in future because of limited reserves and it is estimated that by the year 2050, the 80% of reserve will be exhaused. Also various problems of emission of pollutant gases in environment is another issue.

- The other conventional resource like nuclear fuel is costly and big problem is in its safe utilization.

- To overcome the shortage and ecological problem we need to take over the alternate energy resource for use, known as non-conventional energy.

- The reduction in the supply, and rise in the prices after 1973, forced the developing countries to adopt an alternate non-conventional energy technology, as they get replenished again and again and are sustainable.

- These resources are generated continuously and are also known as renewable energy sources, like solar, wind, bio, tidal, etc.

- The similarity between conventional and non-conventional energy sources is that today the sources which are non-conventional may be conventional tomorrow.

- The non-conventional energy sources are very attractive non-a-day.

12.4.2 Advantages of Renewable Energy

1. It is safe, abundant and clean to use when compared to fossil fuels.

2. Multiple forms of renewable energy exist.

3. It provides the foundation for energy independence.

4. Renewable energy is stable.

5. It is a technology instead of a fuel.

12.4.3 Disadvantages of Renewable Energy

1. Not every form of renewable energy is commercially viable.

2. Many forms of renewable energy are location-specific.

3. Many forms of renewable energy require storage capabilities.

4. Pollution is still generated with renewable energy.

5. Renewables often require subsidies to make them affordable.

6. Some forms of renewable energy require a massive amount of space.

12.4.4 Various Renewable Energy Sources

Various renewable energy sources are :

1. Solar energy
2. Wind energy
3. Hydro Power
4. Biofuels
5. Geothermal energy
6. Hydrogen energy
7. Ocean energy
8. Biomass energy

We are going to study few of these renewable energy soures in this chapter.

12.5 SOLAR ENERGY

- The energy that we receive from the sun is called as the solar energy. The energy produced by the sun is the most important source of energy for all life forms on the earth. It is renewable source of energy. Solar energy means capturing the rays of sun and storing its heat. This heat can be converted with the help of solar panels into heat or electrical energy.

- Solar technologies use the solar energy to light homes and streets, to heat the water, to heat the homes and also to produce electricity. The main benefit of solar energy is that it does not produce any pollutants and is one of the cleanest source of energy. Solar energy helps in lessening the green-house effect. Therefore, it is ecologically acceptable source of energy.

12.5.1 Importance of Solar Energy

- Solar energy is very important for survival of life on the earth. Human beings, plants, animals all require solar energy. Plants require solar energy to produce oxygen and to prepare food (i.e. the process of photosynthesis).

- Solar energy is also required to produce both pure and salt water in oceans as it is the only source of melting frozen ice formed on the mountain caps.

- Solar energy is also used to produce electricity which can be used to run various machines.

- The fossil fuels and other gases and oil that are extracted from the mines are non-renewable energy sources. Also they are costly and cause lot of pollution. But solar energy is renewable source of energy which can be used for lots of activities. Also it is available free of cost. As fossil fuels and other oils are soon going to disappear, the solar energy which is abundant is very important and it should be utilized well.

12.5.2 Uses of Solar Energy

- Solar energy in the form of thermal energy can be used for Cooking/Heating, Drying/Timber seasoning, Distillation, Electricity/Power generation, Cooling, Refrigeration, Cold storage, etc. Some of the gadegets and devices working on thermal energy are Solar cookers, Concentrating collectors, Solar hot water systems (Domestic and Industrial), Solar hot air systems, Solar driers, Solar timber kilns, Solar photovoltaic systems, etc.

- Solar powered hot water systems utilize solar energy to heat water.

- Solar chimneys are passive solar ventilation systems.

- Solar energy can be used for making potable, brackish or saline water.

- Food can be cooked, dried or pasteurized using solar energy.

- Space missions by various countries use solar energy to power their spaceships.

- Solar energy is the preferred mode of creating power where the need is temporary, e.g. temporary fairs, mining sites, olympics, etc.

- Solar driers are used to dry vegetables, fish, grapes, tea, tobacco, etc.

- Solar energy can help in producing power for watches and calculators that do not work on batteries.

- Solar energy can also be used to meet our electricity requirements. Solar radiations are converted directly into DC electricity through solar photovoltaic (SPV) cells. This electricity can either be used as it is or it can be stored in the battery. This stored electricity can be used at night. SPV systems can be used for a number of applications such as Domestic lighting, Street lighting, Village electrification, Water pumping, Desalination of salty water, Powering of remote telecommunication repeater stations and railway signals.

- The efforts are being made to make the efficient use of solar energy. This would reduce our dependence on non-renewable sources of energy and make our environment cleaner.

12.5.3 Advantages of Solar Energy

1. Solar energy is the most readily available source of energy.

2. Solar energy system can work independently without any connection. So it can be utilized and installed in remote areas like forests, deserts, mountains, etc. where there is no sign of electricity.

3. It helps in decreasing the harmful gases and does not contribute to acid rains, global warming, forest destruction and other natural disasters.

4. It is very clean and environment friendly source of energy and do not cause pollution.

5. It is non-exhaustible source of energy.

12.5.4 Disadvantages of Solar Energy

1. Solar energy is available only during daytime with clear sky and bright sunshine. So for night use, storage battery is required, which is very costly.

2. Generation of solar energy is restricted in the areas with cloudy climate.

3. The solar collectors, solar panels and solar cells that are used to absorb heat from the sun are very expensive.

4. The solar batteries that are to be charged are very heavy and require large storage space. Replacing it is also difficult.

5. Installation of solar energy plants requires large area so that system can provide good amount of electricity. This is the great disadvantage in places where the area is small.

12.5.5 Impacts of Solar Energy on Environment

- Solar energy does no produce air or water pollution or greenhouse gases. However, using solar energy may have some indirect negative impacts on the environment.

For example :

1. To make the photovoltaic (PV) cells, that convert sunlight into electricity, some toxic materials and chemicals are used.

2. Some solar thermal systems use potentially hazardous fluids to transfer heat.

3. Large solar power plants can affect the environment near their locations.

4. Some power plants require water for cleaning solar collectors and concentrators or for cooling turbine generators. Using large amount of ground water or surface water in some arid locations may affect the ecosystems that depends on these water resources.

5. Clearing land for construction and placement of the power plant may have long-term effects on habitat areas for native plants and animals.

6. Also the beam of sunlight created by solar power tower can kill birds and insects that fly into the beam.

12.6 WIND ENERGY

- Wind, that is moving air, possesses some kinetic energy due to its high speed. Wind is a result of the solar energy, as heating of land results in movement of air. Wind energy can be harnessed and used for generating electricity or for other smaller purposes by a windmill. In olden times, windmills were used to draw water out of wells or to grind flour etc. It is the rotatory motion of the shaft in a windmill that is used to rotate the turbine and convert it into the form of energy we need it in.

- The main advantage of wind energy is that harnessing it doesn't disrupt natural processes or harm the environment, unlike a lot of other energy sources. To generate electricity on a large scale, a number of windmills are set up over a large area, called a wind energy farm. Such areas need a wind speed of 15 kmph.

12.6.1 Advantages of Wind Energy

1. Wind Energy is an inexhaustible source of energy and is virtually a limitless resource.
2. Energy is generated without polluting environment.
3. This source of energy has tremendous potential to generate energy on large scale.
4. Like solar energy and hydropower, wind power taps a natural physical resource.
5. Windmill generators don't emit any emissions that can lead to acid rain or greenhouse effect.
6. Wind Energy can be used directly as mechanical energy.
7. In remote areas, wind turbines can be used as great resource to generate energy.
8. In combination with Solar Energy they can be used to provide reliable as well as steady supply of electricity.
9. Land around wind turbines can be used for other uses, e.g. Farming.

12.6.2 Disadvantages of Wind Energy

1. Wind energy requires expensive storage during peak production time.
2. It is unreliable energy source as winds are uncertain and unpredictable.
3. There is visual and aesthetic impact on region.
4. Requires large open areas for setting up wind farms.
5. Noise pollution problem is usually associated with wind mills.
6. Wind energy can be harnessed only in those areas where wind is strong enough and weather is windy for most parts of the year.
7. Usually places, where wind power set-up is situated, are away from the places where demand of electricity is there. Transmission from such places increases cost of electricity.
8. The average efficiency of wind turbine is very less as compared to fossil fuel power plants. We might require many wind turbines to produce similar impact.
9. It can be a threat to wildlife. Birds do get killed or injured when they fly into turbines.
10. Maintenance cost of wind turbines is high as they have mechanical parts which undergo wear and tear over the time.

    Even though there are advantages of wind energy, the limitations make it extremely difficult for it to be harnessed and prove to be a setback.

12.7 TIDAL ENERGY

Tides :

- Twice a day, every day since time began, the oceans experience a natural event that moves massive amounts of water in a pattern so reliable that it can be predicted months, even years, in advance. This natural event results in the rise and fall of ocean waters that we call the tides.

- The tides occur because the earth and the moon are attracted to each other through the pull of gravity. The gravitational pull of the moon causes the ocean to bulge out towards the moon, pulling it to its highest level, or high tide. The earth is also pulled towards the moon, but with less strength. This pulls the earth away from the water on the far side of the earth, so high tide occurs on both sides of the planet at the same time. Because the earth rotates, we experience two tides per day.

- The power of this massive movement of water is harnessed and converted into useful energy, such as electricity.

Tidal energy:

- It is the energy obtained from the rise and fall of tides. As the tides rise and fall, a massive amount of water moves towards and then away from shore. Turbines placed in the path of this moving water spin as the water passes by. These spinning turbines are connected to generators that create electricity.

- One way tidal energy is captured is with the use of **tidal turbines**. Tidal turbines look like and work like underwater windmills. They utilize turbines with short but strong blades that spin as the tides move and then transmit their energy to an electricity generator.

- Another way tidal energy is captured is with the use of **tidal barrages**. Tidal barrages are special dams that take advantage of the difference in height between low and high tides. Tidal barrages are built across an estuary or bay. When the tide comes in and the sea level rises, water passes through the dam and becomes trapped in a basin. When the tide goes out, gates within the dam release the water, allowing it to flow through turbines that spin and transfer energy to electric generators.

12.7.1 Advantages of Tidal Energy

1. Tidal energy is a cleaner and safer form of energy generation. No polluting smokes or greenhouse gases (such as CO_2) are released into the atmosphere when tidal energy is harnessed.

2. Tidal energy is a natural form of energy which harnesses the power of the earth's natural resources. No expensive mining or extraction equipment is needed to harness tidal energy.

3. We can always rely on the tides to keep moving, so tidal energy is a very reliable source of energy.

4. This is a sustainable energy source. Harnessing tidal energy does not deplete the tides. The tides cannot be 'used up'. Hence, we can harness their kinetic energy without depleting them.

5. Tidal power enables communities and nations to generate their own energy domestically from their own natural resources, without having to rely on energy supplies from other countries.

6. Tidal energy is very energy-efficient – little energy is lost per unit as the kinetic energy of the tide is converted into electrical energy. It has what is known as a 'high energy density'.

7. The electrical energy generated by the tides can be stored for future use.

8. Tidal power generators can be placed not only along the shore of the ocean but also along the lengths of rivers and in estuaries.

12.7.2 Disadvantages of Tidal Energy

1. Unlike coal, oil or natural gas, tidal energy sources cannot be easily transported for long distances.

2. Tidal power generators cost a significant amount to install.

3. In the case of the sea, tides happen only twice every day. So, the tidal energy will only be generated in these two relatively short periods of time each day.

4. Energy generation can get disrupted by extreme weather events. For example, freezing over of seas/rivers, or a freak storm can disrupt or cut off tidally generated electricity.

5. Installing tidal power plants can disrupt the habitats of plants, fish, water birds and water mammals.

6. Though waves are the source of tidal energy, large waves can also cause great damage to tidal power plants.

7. Tidal energy is only suitable for communities that live within easy reach of a tidal body of water. It cannot be installed at the middle of plains or deserts. It can be installed only near the seas, oceans, and rivers.

8. Tidal barrages work like dams to block up the flow of water at the mouth of a river or along the shore. This can have adverse effects on a community's water supply.

9. Some types of equipments used for harnessing tidal energy can damage or disrupt the habitats of important species.

12.8 GEOTHERMAL ENERGY

- The energy obtained from the earth(geo) from the hot rocks present inside the earth is called Geothermal Energy. It is produced due to the fission of radioactive materials in the earth's core. Some places inside the earth become very hot. These are called hot spots. They cause water deep inside the earth to form steam. As more steam is formed, it gets compressed at high pressure and comes out in the form of hot springs which produce geothermal power.

- To harness this geothermal energy, two holes are dug deep into the earth. Cold water is pumped through the first one and steam comes out through the second long pipe which helps in generating electricity. The holes dug for harnessing geothermal energy result in lesser emission of greenhouse gases than due to burning of fossil fuels. Thus if used at a larger scale and more efficiently, it gives a hope to reduce global warming.

12.8.1 Advantages of Geothermal Energy

1. Geothermal energy is both renewable and sustainable. The thermal resources of the Earth will never run out and will be around for as long as the Earth is inhabitable.

2. Geothermal energy is a non-polluting and environmental friendly energy source when compared with fossil fuel alternatives such as coal, oil and gas. The process of producing electricity from geothermal energy has a much lower impact on the environment.

3. There is no wastage or generation of by-products.

5. Geothermal energy can be used directly. In ancient times, people used this source of energy for heating homes, cooking, etc.

4. Geothermal power plants produce large amounts of energy.

6. Maintenance cost of geothermal power plants is very less.

7. These plants don't occupy too much space and thus help in protecting natural environment.

8. Geothermal power plants don't use fuel, so they don't have to rely on fuel prices and they can offer their consumers stable electricity costs.

9. Unlike solar energy, it is not dependent on the weather conditions.

12.8.2 Disadvantages of Geothermal Energy

1. Geothermal power plants cannot be built just anywhere. Only few sites have the potential of Geothermal Energy. Most of the sites, where geothermal energy is produced, are far from markets or cities, where it needs to be consumed.

2. Installation cost of geothermal power plant is very high.

3. Total generation potential of this source is too small.

4. There is always a danger of eruption of volcano.

5. There is no guarantee that the amount of energy which is produced will justify the capital expenditure and operation costs.

6. It may release some harmful, poisonous gases that can escape through the holes drilled during construction.

12.9 BIOMASS ENERGY

- The Biomass is an organic matter that can be used as a source of energy. It is derived from the energy crops to agricultural residues and waste and from the living organisms. Examples of biomass sources are wood products, dried vegetarians, crop residues, aquatic plants, and much more. These are used to create a biofuel. Biofuel, however, is categorized as first generation or second generation. Examples of first generation biofuel are fuels that come from sugarcane and cornstarch while second generation fuels are fuels that come from the sources of agriculture and municipal waste.

- The heat produced by these fuels undergo combustion which is now ready to provide energy for heating and cooking, of electricity, chemicals, and liquid fuels. About 14 percent of the energy supply in the world comes from the use of biomass energy. Some developing countries have been using biomass energy such as Kenya (75%), India (50%), China (33%), and Brazil (25%). Aside from that, industrialized countries also make use of it as an energy source, like Finland (18%), Ireland (16%), Sweden (9%) and USA (3%).

- Biomass is a carbon-based mixture of organic molecules containing hydrogen, oxygen, nitrogen and other small quantities of other atoms such as alkali, alkaline earth, and heavy metals. The conversion of the Biomass to biofuel is achieved by different methods such as thermal, chemical, and biochemical conversions. In thermal conversion, heat is used as a dominant mechanism to convert into another chemical form. In chemical conversion, the conversion of biomass into another form is by using the coal-based processes such as Fischer-Tropsch synthesis, Methanol production, Olefins, and chemical or fuel feedstocks. On the other hand, the biochemical conversion makes use of the enzymes of bacteria and other micro-organisms to break down the biomass.

12.9.1 Advantages of Biomass Energy

1. It is a renewable source of energy.

2. It is a comparatively lesser pollution generating energy.

3. Biomass energy helps in cleanliness in villages and cities.

4. It provides manure for the agriculture and gardens.

5. There is tremendous potential to generate biogas energy.

6. Biomass energy is relatively cheaper and reliable.

7. It can be generated from everyday human and animal wastes, vegetable and agriculture left-over, etc.

8. Recycling of waste reduces pollution and spread of diseases.

9. Heat energy that one gets from biogas is 3.5 times the heat from burning wood.

10. Because of more heat produced, the time required for cooking is lesser.

11. Pressure on the surrounding forest and scrubs can be reduced when biogas is used as cooking fuel.

12. It is a more cost effective means of acquiring energy as compared to oil supplies. As oil supplies are getting depleted day by day, it is becoming a costly commodity.

13. Growing biomass crops use up carbon dioxide and produces oxygen.

12.9.2 Disadvantages of Biomass Energy

1. Cost of construction of biogas plant is high, so only rich people can use it.
2. Continuous supply of biomass is required to generate biomass energy.
3. Some people don't like to cook food on biogas produced from sewage waste.
4. Biogas plant requires space and produces dirty smell.
5. Due to improper construction many biogas plants are working inefficiently.
6. It is difficult to store biogas in cylinders.
7. Transportation of biogas through pipe over long distances is difficult.
8. Many easily grown grains like corn, wheat are being used to make ethanol. This can have bad consequences if too much of food crop is diverted for use as fuel.
9. Crops which are used to produce biomass energy are seasonal and are not available over whole year.

SUMMARY

- Energy is required to do any kind of work.
- Energy sources which cannot be used again after their consumption are called non-renewable energy sources. e.g. Coal, petroleum, natural gas, nuclear power.
- Energy sources which can be used again and again, and stock of energy is not affected by the rate of their consumption are called renewable energy sources. e.g. Solar energy, wind energy, biomass energy, tidal energy, geothermal energy etc.
- The energy consumption of the world has been increasing with industrial growth. The fossil fuel is going to be exhausted in future because of limited reserves and it is estimated that by the year 2050, the 80% of reserves will be exhausted. Also problems of emission of pollutant gases in environment is another issue. Due to this it has become necessary to use the non-conventional sources of energy.
- There are advantages as well as disadvantages of the non-conventional energy sources. But still they are very useful to overcome the energy crisis to be faced by the world today.

EXERCISE

1. Define 'Non-renewable energy sources'. Give examples of it.
2. Give advantages of non-renewable energy.
3. What are the disadvantages of non-renewable energy ?
4. Write about the necessity of renewable energy sources.
5. Write advantages and disadvantages of renewable energy.
6. List various renewable energy sources. Explain any one of them in detail.
7. What are the impacts of solar energy on environment ?
8. Write short notes on :
   (a) Wind energy.
   (b) Tidal energy.
   (c) Geothermal energy.
   (d) Biomass energy.
9. Give any four advantages and disadvantages of following energy sources :
   (a) Solar energy.
   (b) Wind energy.
   (c) Tidal energy.
   (d) Geothermal energy.
   (e) Biomass energy.

□□□

1. **Choose the correct answer :** $(2 \times 10 = 20)$

   (i) Dispersion is maximum for ......

      (a) Violet colour

      (b) Yellow colour

      (c) Red colour

      (d) None of the above

   (ii) The proportionality constant in Coulomb's law has unit of ......

      (a) Farad

      (b) Farad/meter

      (c) Meter/Farad

      (d) Newton

   (iii) Potential has the unit of ......

      (a) Joules/Coulomb

      (b) Joules

      (c) Joules/m^3

      (d) Joules/m^2

   (iv) Ohm's law is valid for ......

      (a) Electrolytes

      (b) Conductor

      (c) Semiconductor

      (d) None of the above

   (v) In an optical fiber, light propagates in the ......

      (a) Core

      (b) Jacket

      (c) Cladding

      (d) Air

   (vi) When reverse bias is applied to a P-N junction diode, it ......

      (a) Lowers the potential barriers

      (b) Raises the potential barriers

      (c) Greatly increases the majority-carrier current

      (d) Rate of thermal generation of electron-hole pairs.

   (vii) In photoelectric effect, emission of electrons is due to ......

      (a) Electric field

      (b) Thermal energy

      (c) Electromagnetic radiation

      (d) Applied potential difference

   (viii) Lasers are used in ......

      (a) Metal cutting

      (b) Welding

      (c) Communication

      (d) All of the above

    (ix)   Which of the following wavelength falls in X-ray regions ?

        (a) 10^{-4} A°

        (b)   1 A°

        (c)   1000 A°

        (d)   10000 A°

    (x)   Primary source of energy is ......

        (a)   Sun

        (b)   Water

        (c)   Coal

        (d)   None of the above

2. State and explain Snell's law of refraction of light. Define refractive index. **(4)**

    Ans. Refer to Section 1.3 on page 1.3.

3. Define :

    (a)   Photoelectric effect.

    Ans. Refer to Section 8.1 on page 8.1.

    (b)   Stopping potential.

    (c)   Threshold frequency.

    (d)   Work function.

    Ans. Refer to Section 8.6 on page 8.3.

4. Define X-rays. Write four properties of X-rays. **(4)**

    Ans. Refer to Section 10.1 on page 10.1. Also refer to Section 10.4 on page 10.5.

5. Define conventional and non-conventional sources of energy. **(4)**

    Ans. Refer to Section 12.3 on page 12.3 and Section 12.4 on page 12.5.

6. Write four properties of Laser. **(2)**

    Ans. Refer to Section 9.9 on page 9.6.

7. Define electric lines of forces and write down any four properties of electric lines of forces. **(6)**

    Ans. Refer to Section 2.5 on page 2.3.

8. Obtain an expression for the potential at a point due to a point charge using integration method. **(6)**

    Ans. Refer to Section 3.3 on page 3.3.

9. Derive an expression for the capacity of a parallel-plate condenser. **(6)**

    Ans. Refer to Section 4.4 on page 4.3.

10. Apply Kirchhoff's law to derive the balanced condition of the Wheatsone bridge. **(6)**

    Ans. Refer to Section 5.10 on page 5.8.

11. Derive an expression for acceptance angle for an optical fiber. How it is related to numerical aperture ? **(6)**

    Ans. Refer to Section 6.7 on page 6.5.

12. What are p-type and n-type semiconductors ? With suitable diagram, explain the working of p-n junction diode in forward biased condition.

    Ans. Refer to Section 7.3.2 on page 7.6. Also refer to Section 7.4 on page 7.8.

13. Define :

    (a)   Nanoscale

    (b)   Nanometer

    (c)   Nanoparticle

    Write two applications of nanotechnology.

    Ans. Refer to Section 11.2 on page 11.1. Also refer to Section 11.3 on page 11.2.

•••